T0256193

Land Sickness

Land Sickness

Nikolaj Schultz

polity

First published in 2023 by Polity Press

Polity Press
65 Bridge Street
Cambridge CB2 1UR, UK

Polity Press
111 River Street
Hoboken, NJ 07030, USA

ISBN-13: 978-1-5095-5612-0
ISBN-13: 978-1-5095-5613-7 (pb)

A catalogue record for this book is available from the British Library.

Library of Congress Control Number: 2022943243

Typeset in 12.5 on 15pt Adobe Garamond Pro
by Cheshire Typesetting Ltd, Cuddington, Cheshire

The publisher has used its best endeavours to ensure that the URLs for external
websites referred to in this book are correct and active at the time of going to
press. However, the publisher has no responsibility for the websites and can make
no guarantee that a site will remain live or that the content is or will remain
appropriate.

Every effort has been made to trace all copyright holders, but if any have been
overlooked the publisher will be pleased to include any necessary credits in any
subsequent reprint or edition.

For further information on Polity, visit our website:
politybooks.com

Contents

Problems

It never stops. The problems never seem to leave me alone, they follow me around all day, from morning to evening, from sunrise to sunset. They have been doing so for a long time, but tonight is different. Now, they even follow me into my dreams.

I have been going to bed late for a while, not by choice, but because the heat in this city is unbearable. The warmth incapacitates my body and mind: everything seems slower, each minute longer, every movement heavier. There is yet another heatwave in Paris, one of those that used to be unusual but have come to seem normal, or at least familiar. The heat exhausts me. I am tired and need to sleep, but as I close my eyes, my heartbeat speeds up. A tingle runs through my arms into my fingers, as my chest tightens and my neck stiffens. I am not sure which came first, the feeling or the thought, but I know this: the problems have caught up with me. The breeze that was supposed to calm me down has triggered the alarm bell. The fan I cannot sleep without turbocharges my energy consumption, emitting more CO_2 into the atmosphere, resulting in yet more heat. Cooling my body down has its price: a cost probably first and most violently paid by somebody else, most likely somewhere in the Global South.

I turn over and glance out through the small crack between my curtains. How do the couple living across the road, under the metal roof, manage? Their flat is small, even smaller than mine, with only one window, making it almost impossible for the air to circulate, and I doubt if they have air conditioning. They must be suffocating in there.

It seems the Anthropocene is not a nice place to sleep. Even if I am exhausted, I suppose I could make the most of it and work, as I normally do when sleepless, but as I roll over and prepare to get up, I remember that this part of my life too has become haunted. Yesterday morning, I realized that what I always wished for – having my name on the cover of a book, displayed in some Parisian bookstore – sends me hundreds of miles away and plants my feet straight in some ancient woods, contributing to deforestation. With each word printed on paper in black ink transported from afar, volatile organic compounds are released into the atmosphere.[1] With every single page of the book I write, the more deeply intertwined I become with the issues; maybe just a little, but enough to be an active participant in the unfolding of a planetary emergency.

Sirens grow louder and then disappear as an ambulance races along the street below. I am now wide awake. Sitting up in bed, hands locked between my thighs, it seems that the only stable structure left is the wall supporting my back. I lack a sense of direction and have lost my bearings: behind me, in front of me, alongside me, up in the sky and below my feet, I see nothing but signs of this mess. Wherever I fix my gaze or turn my imagination, I recognize the disturbing trails of my own being, its activities and doings. My sight blurs in the dark and I begin to feel short of breath. These problems dog my footsteps; no, worse, they *are* my footsteps. How can I aspire to anything when all possible points of view remind me of the points of life I disturb? How can I dream at night when my last thoughts are moral vertigo over the cost of my sleep? How can I dream during the day if what wakes me in the morning further implicates me in the catastrophe?

These issues do not go away. They accompany me every morning when I read the newspaper and learn about yet another manifestation of climate change. This afternoon, it might have been 43° in Paris, but in California it was 54°. In Central Europe, hundreds of people are dead because of flooding, and in Indonesia dozens are still missing after a cyclone struck a few months ago. The problems fill up my basket when I do my grocery shopping at the supermarket, item by item, each one wrapped in plastic that ends up in the ocean somewhere. I stopped eating meat, but the avocados and quinoa I opted for instead cause soil degradation and water shortage where they are grown. In the mornings, the coffee I need to wake my mind destroys soils and discharges waste and pollution into foreign rivers. The problems even splash onto my shoulders when I take a shower, with each minute under the streaming water releasing more CO_2 into the atmosphere. But when I step out and dry myself off, new problems accumulate with every piece of clothing I put on, products of an industry hugely responsible for the global emissions of greenhouse gases. These issues are not easy to digest, they cannot be washed away or covered up. Every day, I learn how another aspect of my life is interpenetrated with these problems, and how my being in the world entangles me in these troubles.[2] Every day, I realize that the problem is *me*.

Beings

I barely recognize myself. The image of the human species seems to have been transformed,[1] and I am mutating along with it. I feel this transformation every day, the change inside and out, as most of my personal actions or habits mirror these global patterns. Like some planetary Faustian bargainer, holed up in the dark of my bedroom, I am paying the ethical cost of my own affluence. Silently lying in bed, I hear the echoes of a thousand remote voices. Almost no matter what I do, the extensions of my livelihood pull traces after me, leading directly back to distant calamities. The material footprints of my life always situate me elsewhere, far away, most often a place where my presence is hazardous.

I am here, there and everywhere, and this is why the problems never disappear. A colleague recently found a way to frame this dispersion[2] by distinguishing between the world I live *in* and the world I live *off*. My livelihood was always somewhere else, as the threads weaving my life together never fail to displace my being in this world. For me to live and thrive, I have to set foot on the lands of others. Yet I still do not have a language that captures the *experience* of me breathing heavily in my bedroom while my boots are buried deep in someone else's livelihoods. In high school, I learned that when meeting a person, a bit of their destiny was in my hands. This is different: now, what I do has effects in places where I have never been and probably never thought of visiting; it concerns people I have never met, and whose lives I can only vaguely imagine. However, there I am, in the midst of their livelihoods, affecting their possibility of eating, drinking, breathing and living. And it is all due to my being, my freedom and my own way of life. My life at the expense of theirs, on a summer night during this heatwave.

The temperature does not drop, but I feel frozen, fixed, immobile, afraid of my every movement. Somebody seems to have installed Chateaubriand's Chinese Mandarin button underneath my feet,[3] and it is triggered by each step I take. As I grope for solutions, it strikes me that if this is suffocating and confusing, it is probably because it reshuffles certain existential clues, forcing me to reconsider them anew. Shoulders hunched, I crawl out of bed and begin browsing my bookshelves until I find first the book, then the page, I was looking for:

> This is what I need, and this is what I strive for. A man must first learn to know himself before knowing anything else. Not until he has inwardly understood himself and then sees the course he is to take does life gain peace and meaning.[4]

Bewildered, I swiftly close the book, as if it has done me more harm than good, aware that I can probably never open it up in the same way again. Being Danish, I know this page of Kierkegaard by heart, since it used to offer me guidance when asking myself what something was and was meant to be. Yet tonight it no longer resonates: travelling deeply *inwards* does not assuage the dread and anxiety that beset me since they are caused instead by the *outward* traces my being is leaving behind. It is as if my existence has flipped to the outside, and so I flip the pages in turn, but in fact nothing in this book equips me to frame, phrase or fathom the contours of my being. In the dark, my silhouette looks the same, but the shadows it casts are different. My very existence in the world has changed, and trying to reassemble its different pieces puzzles me, to say the least.

If I cannot sleep, it is because I have been transformed into a weird monstrosity of a species that I do not particularly like, that my mind hectically tries to grasp, but I lack the words and expressions to understand. The insights from the existentialist tradition I was fond of as a teenager no longer hit the spot. They fail to describe or explain the being I am tonight, and will continue being tomorrow when the sun rises again. Existence probably still precedes essence,[5] but this new existence is definitely another sort of being, one that is constantly fleeing home. It is not just that I exist *for myself*, as if I was lodging in some private hotel room full of mirrors. Rather, it seems that I exist *from others*, like a spider in a web, sustaining myself by catching and feeding off them. As I weave my silken threads, my being and its trails constantly borrow from, overlap with and obstruct the continuous being of other entities: some of them close, some of them far; some of them human, some of them non-human. *Intermixti, ergo sum*: I mix and interfere, therefore I am, and continue to be. I am living in a door-less house, and beyond my exhausted limbs my being extends, leaves traces, sets traps and forms trails that other beings are forced to travel by. It is not merely that my existence unfolds as a continuous making of meaning. Rather, it appears to *fold* as a process of negotiation between me, the entities that allow me to live and the beings that my existence touches and is entangled with.[6]

Yes, if these walls could speak, they would probably tell me how I have become a peculiar sort of broker, settling business where I have none, clinching deals where I should not, negotiating where I have no diplomatic rights. At least that is how it feels. My being may unravel as a construction of identity, but inseparably so from the destruction of another entity. I am earth, wind, fire and water, perhaps, but I am also soil depletion, hurricanes, wildfires and sea pollution. The nausea that keeps me from falling asleep is due not to the eternal, inescapable depths of my freedom or endless corridors of my own subjectivity,[7] but to its recurrent, external traces in the world. Here lies my bad faith, not in social conformity or subjective inauthenticity, but in the immeasurable, destructive social and natural vestiges that my actions entail. This is what seems impossible to escape – not freedom as such, but the material cost of it.

Like a bedridden patient, obsessed with understanding his own diagnosis, my condition remains a riddle to me. Yet the more I dwell on the topic, the more it seems the fundamentals of the existentialist question have changed shape. What worries me is still 'being', but the question is no longer the *meaning* or *purpose* of humans and their existence, but rather which beings *allow* other beings to remain in existence or *prevent* them from doing so. Not only is my existence thrown into a world,[8] its vestiges are first tossed all around it before boomeranging back to me.

Years ago, I spent sleepless nights searching for the meaning of my unique human condition. Tonight, instead, my fear sends me looking for the multiple sorts of earthly, material beings that condition human existence, that humans always co-exist with, at a moment where such conditions for life can no longer be taken for granted precisely because of humans' way of being.

As I notice how my sweating body is sticking more and more intimately to the bedsheets, my mind flips between the two sides of this newly tossed existentialist coin. On the one side is the realization of the many different beings and existences I depend on in order to be. On the other side is the understanding that I, a human relying on these entities, am now engaging in the annihilation of such beings – and therefore the livelihoods of other humans. This is my new human state, penetrated by the planetary scale of my species' actions, converting it into its terrestrial condition. Existence has become a lot more complicated, unfolding in a new geography blurring the local and global, and including multiple sorts of beings, their appearances and disappearances.[9] The existential task today is how to situate oneself in relation to these processes.

In more than one way, the question of existence has become decentralized; yet, sitting up in bed, I remain at the very centre of it.

Generations

As the clock strikes midnight, my phone notifies me that tomorrow is my grandmother's birthday. Normally, I would call her; among the cleverest people I know and typically awake at this hour, she could help me get out of a shambles like this. She has always known how to wrap her head around problems, and since I was a boy, she has often accompanied me when I have wondered what it means to be, either alone or together. A compass for my life, she has offered me clues and directions, and, as part of the post-war generation who built the economy I benefited from and who saw it as my destiny to contribute to, she has embodied where I have come from and am going.

Yes, usually she would be able to advise me, but

she too has changed. For the same reasons, but in a different manner, she too has been transformed into another kind of human, with another present, past and destiny. She too has mutated into an unfamiliar being with a tail wagging behind her, and with a responsibility none of us knew she carried until recently.

I could call her, but when we were last together, we could not speak. My mind travels back to that time, Christmas Day, in her home in the south of Denmark. All together preparing lunch in the kitchen, festive songs playing in the background, the seasonal peace suddenly melted as my brother reminded us how it used to snow at this time of year when we were children. His remark was certainly no battlecry, but it immediately had my grandmother transfixed. From speaking of everything and nothing, she grew silent, turned her back and withdrew from the conversation.

My grandmother would never deny the ecological catastrophe, but neither could she speak of it, and I understood why. Her silence reflected the battles she and her generation fought, the narratives that made them bearable, and the values making them necessary. The struggles central to her political biography rested on a moral and historical certainty, with history unfolding as a linear process of progress towards freedom and affluence. All she struggled for was for these values to be realized, guarded and inherited by her children and theirs. This was her political horizon, a heritage proudly offered to her descendants, one they would embrace with gratitude and confidence.

Now, to her horror, things had changed, attaching another meaning to her biography. Not only did her grandchildren no longer believe in the sanctity of her horizon. That would have been tough enough: to realize by the end of your life that your beliefs were outmoded or redundant. But this was worse: she realized that it was the very horizon that she had fought for that had mired her descendants in danger and disaster. No, her generation had neither guided their successors safely into the future nor opened up fruitful avenues for them. Instead, they had taken away their earth, their horizons and their possibilities of life. They had not secured their prosperity, but had burdened them with a debt to be paid. They had freed themselves, but at the cost of taking their successors hostage.

As my body wrestles with the bedsheets, I realize my future is buried in my grandmother's past.

The spatial disconnection between the world one lives *in* and the world one lives *off* was always associated with a temporal disconnection between the time one lives *in* and the time one lives *off*. My grandmother's generation lived *in* the present but *off* the future, something made visible today with the endangerment of future generations' material conditions of habitability. The 'generational issue' again lands at the core of political conflicts, but this time in a shape closer to its etymological roots, since it has become inseparable from the *genesis* of living conditions. Unfolding along a new geo-historical temporality, it now entangles with land, soil and nature. Woven together with a wider set of earthly processes, it no longer suffices to define generations in terms of cohorts of people of a similar age, organized around a shared experience of historical-cultural events.[1] Instead, what *ties or unties the knots* between generations, what unites or separates them from each other, is the different relationships and temporal horizons of their territorial conditions of liveability: if they had, have or will have such means of subsistence; if they have had them stolen; and what the prospects are of giving or getting them back. When young activists frame their struggles in terms of a 'generational battle', this makes sense on more than one level, since at the heart of their enterprise is the genesis of the planet's conditions of life, whom and where they derive from, what is left behind for them, and how to manage their lives from there.

If my grandmother was silent, it was because she was the protagonist of a generational conflict centring on the colonization of territory *and* time, a conquest attacked by those who had had their soils and futures stolen, with the aim of superimposing both the worlds and times you live *in*, and those you live *off*. Everyone is entitled to inhabit their own land and their own time, the young rightfully cry, and if this claim is so explosive, it is because it involves suspicion of the passage of time itself. It attacks the certitude of the generational logic of automatic temporal succession. 'Fridays for Future' is a declaration of war against those who took the future for granted and were certain of the direction of travel towards it. 'How dare you!' is the accusation levelled at those who were so sure about history's trajectory that they ignored all the alarm bells.

Living among the ruins of such certainties, the new generation demands another reflexivity of time and space, devoid of any certitude about how and when the wind blows, with the aim of temporal and territorial justice. My grandmother felt the weight of these injustices, because no matter what my brother's intentions were, his remark reflected their realities. How could she possibly speak? How could she ever offer solace or advice? The facts could not be spoken. They were to be silenced. Not because they were false; on the contrary, it was because their truth was carving her up from inside and out. As they did so, I could only try to balance between empathy and anger. Individually, she too is victim of a planetary metastasis. Collectively, I blame her and her generation for it.

Christmas used to be something else, the atmosphere different. There is not a snowflake in sight. How can you wake up in the morning knowing all your horizons have become obsolete? How can you dream at night knowing all you struggled for has led your loved ones to a dead end? These problems follow me around from dawn to dusk, invade close to all aspects of my life, changing my being and the ground beneath my feet. They travel through time and cut through generations, making warzones out of dinner tables at family reunions. Not only do they forbid me from dreaming, they even take hostage of my loved ones, transforming them into villains. They do not leave me alone, and neither do they spare those I care for, and those who care for me. In their eyes, I have changed, as they have changed in mine: a cross-generational mutation, with both points of view suddenly transformed, scattered as we have become along more or less privileged points of life. These issues force you to judge others as you judge yourself, shining another light on your past and future, demanding that you re-evaluate your predecessors, contemporaries and successors. Indeed, it truly is very Kafkaesque: first *Metamorphosis*, then *The Trial*.[2]

Transmissions

The sun is rising, and with it, once again, the heat. My request is the opposite of Diogenes': all I ask for is a bit of shade, but the sunlight is already burning through the curtains. Although I've been going to bed late for a long time now, that was the first night in a while I didn't sleep a wink. Drowsily, I manage to get to the bathroom to splash some water on my face, but returning to the bedroom, it feels like a prison. These torments won't leave me alone. I want to run away, to break free, to escape what I have become, what is mine, what my predecessors have become, what all this has become. I want to jettison the traces dragging after my life, to detach from the material consequences of my existence, to be isolated. Yes, I want to be an island.

Suddenly it hits me: I know what to do and where to go. For the last week or so, my friend Victor has tried to convince me to take some time off and join him on his family's sailboat. Usually, the boat docks in Marseille, but currently it is anchored on the French island of Porquerolles. I have heard of it before because of its famous crystal-clear blue sea, its flora, fauna and land-scape. An isolated, paradisiacal island in the middle of the wide, blue Mediterranean Sea: this must be the ideal get-away, the perfect escape. I want to be an island, so, *voilà*, I shall flee to an island.

Eager as an inmate discovering a hole in a jail wall, I begin preparing my escape immediately. I must pack at least a few essentials, but as I dig through my closet, obstacles appear in my hands. The vertigo returns: for each t-shirt I throw in my bag, 2,700 litres of water, 2.6 kilograms of CO_2 emission; a pair of jeans, 3,300 litres of water, 11 kilograms of CO_2 emission.[1] I want to travel light, but the items I bring carry a new weight; no matter how neatly I try to fold these clothes, the mess piles up. I attempt to calm down, telling myself it does not matter, because I have found a way out. I just need to get to the island.

A few minutes later, I close the door behind me, but as I rush down the stairs I remember that my escape has become more difficult due to a certain virus that has brought societies worldwide to a grinding halt. To restrict its spread, politicians, in dialogue with scientists, have decided that I cannot travel without a piece of paper proving that I am not a carrier. Luckily, the pharmacy on the corner can provide me with such a pass in around twenty minutes if I agree to be tested and if the test reveals no signs of infection. The pharmacist begins by inserting a swab deep inside my nostrils and rotating it. Next he places the swab in an extraction buffer tube and rotates it while squeezing the sides between his thumb and forefinger. Then, as he removes the swab, he squeezes the tube again to force the sample from my nostrils into the liquid. Three drops of it are then added to the specimen well of a test device that fifteen minutes later will tell him my result: if coloured lines appear in the upper and lower sections of the results window, I carry the virus. Regardless of the outcome, my results are registered on a computer, added to the national virus database used to take political decisions and connected to a printer that translates the state of my body into a piece of paper, used to decide whether or not I can move around.

My result is negative, so I can travel. In handing me my proof, the pharmacist simultaneously confirms that the maintenance of social order is predicated on the intervention of science and our alliance with the facts that it constructs.[2] Although politics and science might differ in their respective practices and tonalities, their instruments play in the same orchestra, tuning in and out, co-defining each other's soundscape. Science does not reign beyond the realm of society, accessing some universal truth; rather it co-constructs it from inside, both by stirring up controversies *and* by manufacturing somewhat stable facts that change the world, and, in this case, allow me to make my way through it.

As I stride towards the metro, it strikes me that the pharmacy test is another example of the many components that have to be in transmission in order for me to transport myself. The pandemic made us aware of a world inhabited not only by humans, but also by many other entities that live among us, co-defining our lives, some of them being viruses, bacteria and vaccines. For me to remain in motion and reach my destination, multiple existences – among them microbes, a swab, a test device and a computer – all have to interact. I never move disconnected *in* space; rather, I have to act *with* other beings if I want to remain moving, in entanglements and engagements that always redefine space. For me to pass through the world depends on the passage of many others.

A new material reflexivity centred on these principles of agency seemed to spread together with the virus, as it made clear how we are always related to other existences, how we never fail to co-exist and co-act with other beings. It illuminated how all these entities, in their unfolding, drag traces with them that remain in place, constructing or deconstructing spaces. At the Gare de Lyon, signs of this reflexivity are everywhere. We have taught ourselves to wear a mask, keep our distance and stay attentive to the 'protective measures'. Strangely, these artificial micro-isolations have shown us that we always leave traces, that we are always displaced, elsewhere. '*Larvatus prodeo*' was Descartes' motto,[3] but even if you advance masked, your transition is always a transmission.[4]

As I move through the station, the pandemic has made me attentive to the multiple trails I transmit and leave behind, how they modify spaces, how they might obstruct spaces, allowing another, somewhere down the lines of vestiges, to keep breathing, or preventing them from doing so. By a chain of touches, the ticket vending machine might infect me or someone else, and further down the road of overlapping pathways, these traces might obstruct the lungs of a young asthma patient, a homeless person in fragile health or some old man on his way to see his grandchildren. We all leave traces, they follow us in our footsteps, they *are* our footprints, and as they appear, it seems that *we* never simply disappear. My train is leaving the platform, it is taking me south, but as I continue my escape, parts of me are still passing through the north. As the train gains speed and leaves the station, the landscape through the window swiftly alters; the land is changing, and I am changing with it.

Oceans

It is almost night when I arrive at the boat. Before we have dinner, Victor shows me my narrow sleeping cabin and the vessel's interior. We will need to wait for daylight, he tells me, before I can understand the instruments that are so crucial to the boat's operation. As I try to accustom myself to the gentle rocking of the confined space that will shelter me over the next few days, Victor's younger brother, Paul, begins detailing the boat's past. Designed by English naval architects and built in a shipyard in Brittany, the 12-metre-long craft, made of blond mahogany, was bought by their grandfather in the early 1970s, a few years after its first sea launch, and given the name *Timia*. Since then, I learn, it has had quite a history. Over the years, it has won prestigious regattas, been navigated by famous sailors and made numerous passages across the oceans, from the Mediterranean as far as the West Indies. Indeed, *Timia* has witnessed grandeur and happiness, but is also acquainted with tragedy: during a storm while the boat was sailing in the West Indies in the 1980s, its mast was torn off, almost destroying the wooden hull.

As Paul pauses his account of the boat's history, my thoughts drift to another sailor. It was in these vast Mediterranean waters, almost completely enclosed by land, that Jules Verne saw the beginning and end of the globe.[1] In this most powerful of earth's climates, he recalled, through Michelet, how man could renew himself. The haunted Captain Nemo hailed the endless sea, this grand reservoir of nature, but still he felt trapped there, squeezed in between the shores of Africa and Europe. Looking across the sea from the boat's deck, I too sense the beginning and end of the globe mirrored in these waters – not due to their endless circularity, but because I know it was in the labyrinth of this sea that the expansionist maritime powers invented the dream of globalization,[2] a dream whose end I glimpse in the ocean's destructive metamorphoses.

From the stern of the boat, the waters look peaceful tonight, but beneath their surface, they are undergoing a violent transformation. Decades of nearly unregulated industry, production and economic activity have damaged these waters' ecosystems almost beyond repair. In the last fifty years, since *Timia* began sailing around the Mediterranean, it has lost nearly half of its mammal population and a third of its fish species. Its green turtles are in danger and more than half of its sharks are at risk of extinction. During the same period, there has been a more than 30 per cent decline in distribution of the endemic *Posidonia oceanica* seagrass, labelled the lungs of the Mediterranean due to its capacity to produce oxygen and absorb CO_2 from the water, together with the disappearance of over 13 per cent of its anthozoan sea anemones and corals.[3] At the same time as these species die or flee, the sea level here could rise by up to a quarter of a metre in the next thirty years, while human migrants, desperate enough to put their faith in the hands of traffickers, fight against the waves in search of better lives. The damned of the sea,[4] and their many divisions . . . In the ocean, where the horizon of globalization was born, the incompatibility with the planet's conditions of habitability, allowing life to prosper, appears in the waves.

Hence, like Captain Nemo, I too feel confined here in these waters, not due to the shores, but because I am tightly clamped between two conflicting images of the world, each offering different perspectives on where I am, and where I am going.[5] If I feel locked at sea, it is because the endless globe I thought I lived on and had as my horizon no longer seems boundless but has mutated into a finite earth, with a set of planetary boundaries for habitability, some of which my species has already brutally crossed.[6] Beneath the surface of these waves, gently splashing against the bow of the boat, another planet is coming into view, one that might be round like the globe but is far from infinite. In this sea, I do not travel through an unlimited, open space; I pass through a planet remade by the traces of my journey, with earthly constraints emerging as a consequence of my species' movements. The planet and the seas are shrinking in habitability beneath the boat, and another world emerges in its mirror image, one in which I am renewed, along with my species, and transformed into a different kind of figure.

Paul resumes his biographical tale, and I soon begin to understand how this boat is not one but many. So many sailings through different waters, so many sailors on the boat who knew their direction. *Timia* has existed for much longer than I have, and chances are it will still be here after I have gone. Since its maiden voyage, it has seen the waters change, the wind turn and the waves enfold the progress of many storms, their beginnings, middles and ends. Now this vessel carries and contains me, another sailor, disoriented without a fixed course, at a moment when the seas are rising, the waves have become uncontrollable and the direction has become uncertain. A peculiar seafaring explorer, one whose adventure is not to find new lands on the horizon of an endless globe, but who is discovering altered lands beneath his feet, on a planet with boundaries made by and for his kind of being. New lands I still do not know how to travel on or fully grasp myself as a being of.

Here, on this sea, Captain Nemo felt his loss of independence and liberty as his ship was confined between the rocky shores, and yet I have fled here because I feared my freedom had abandoned me. Still, as I manoeuvre myself into the little, lamp-less cabin bed after dinner, I cannot help but think that my escape plan appears to have worked. Since Victor picked me up at the ferry, my mind and body have grown calmer. The closer I have got to the island, I think to myself, the closer I seem to be to consolation. Here on the boat, a few hundred metres from the shore, the breeze is cooling my body and my problems have lessened their pursuit. My ears relax to the sound of the wind mixing with the waves that gently rock the boat. Through the little porthole window, I notice stars in the gaps between the stationary clouds. Tomorrow I will reach my destination, my isolation, my island.

As morning comes, rays of sunshine enter through the same window and wake me up. I slept. My mind feels rested and my body agile as I crawl out of my berth and make my way onto the deck. I certainly did not leave the heat behind when I left the city, but it is freshened by the breeze and no longer suffocating. Scouting across the motionless waters, I am finally able to see the island properly rather than the shadowy outline of the previous night. I only get to admire it briefly, however, before Victor joins me on deck. Almost immediately, he begins guiding me through the technicalities of the boat, briskly jumping from ropes to knots, from sails to how to fold them, to the anchor and its weight. To my father's disappointment, I was always a practical disaster, and despite Victor's best efforts, I find it difficult to follow and to concentrate. I know the look on his face because I saw it throughout my childhood. Victor has discovered my lack of focus and enthusiasm for the mechanics of the boat, and after a few minutes he gives up. Sensing that my mind has another goal, he suggests we postpone the technical details until later, and offers to drop me on the island straight away.

Islands

As the largest of four islands constituting the Hyères archipelago, Porquerolles stretches about 7 kilometres in length, 3 kilometres in width, amounting to a total area of 12.5 square kilometres of land. Guarded on the south coast by steep limestone cliffs, and on the north coast by beaches and a port, the island's hills seem to float like clouds above the water. As we approach the harbour, I glimpse between the hills the fortresses a young Captain Napoléon Bonaparte built after English troops ravaged the island at the end of the eighteenth century. On top of its southern hills, the lighthouse has been greeting sailors since 1837, some twenty years after the island's village was first established. At the beginning of the twentieth century, a company invested in Porquerolles' infrastructure, set up a power station, constructed water retention basins, pipelines and more, before selling it to a rich engineer named François Fournier[1] in 1912, who bought it for his wife, with the plan of making the remote and isolated island self-sustaining. Over half a century later, in 1971, the French state purchased vast stretches of it with the aim of preserving its nature and wildlife, a responsibility assigned to Port-Cros National Park, which eventually received ownership of over 80 per cent of the island.

A distant island built and developed with self-sufficiency and the protection of untouched wilderness and nature in mind. Isolation, untouched spaces and a lack of traces: as the boat slows on its approach to the port, all of this makes me think that the island will be a perfect bolthole for freedom, or at least a place I can escape the problems for a while. Stepping onto the harbour bridge and watching Victor disappear out of sight, I briefly consider strolling around the port village, but nothing seems open, and the port is close to deserted. Realizing the village has not yet woken up, I instead decide to head directly towards the little hidden piece of beach Victor and Paul talked about last night on the boat. Sitting down on a wooden bench facing the coastline, I unfold an old map of the island Victor lent me. Later, perhaps, I will be able to rely on my memory, but for the time being, I need a paper tool like this to get a sense of place. Struggling at first to find my location vis-à-vis my destination, I soon begin to understand the map's reference points. I need to head a few kilometres east, along a small road running parallel to the rocky shoreline. Following another brief moment of confusion, I seem to find my way. The road's curves, openings and dead ends all start to arrive where and when I expect them to, they match their depiction, and so, by referring back and forth between the map's coordinates and the route in front of me, I manage to find the right path.

After passing through a small neighbourhood of white and yellow stone houses, I find myself entering an unpaved gravel road, bordered by agricultural fields on the one side and forest shrubbery on the other. Small stones and dust particles whirl into the air every time bicycles pass me, stinging my eyes, and so, after a hundred metres or so, I am relieved when the map indicates I have to take a full 90 degree turn and pass through a forest trail opening up in front of me on the right. With the gravel road now behind me, green and brown plants, branches, leaves and bark fill up the forest floor, and as the sea begins peeping out between trees and bushes, the trail starts heading downward.

From the glimpses flashing through the branches, I anticipate I will soon have to clamber down the cliffs, but as I reach the end of the trail, I see that a slope has been cleared. Meandering my way down it, I glimpse the site Victor showed me on his phone, and as I enter the sand of the beach, relief spreads through my body. Only a few more steps and I am there: the flawless stretch of beach on the ideal island, the ultimate isolation and the perfect escape, as distanced from my problems as I could ever have imagined and hoped for.

At least I thought so. But less than a minute after I throw myself and my stuff onto the sand, a desperate old woman approaches and reminds me that today not even an island is isolated.

'I'm sorry to disturb you but can't you please go somewhere else? I've been walking in the sun all morning, but I can't find any place to lie down or rest. I have lived on this island all my life, but now there is no more space for me. I cannot find a single metre of beach left for me. I was born here, I am *Porquerollaise*, but I no longer know where to go.'

My French is not great, but I cannot pretend not to understand her. Immediately feeling my pulse racing, not knowing how or what to respond, I pick up my belongings and begin stumbling backwards in the sand, back the way I came from. She is one of the island's couple of hundred permanent inhabitants, one of the few born here. This is her dwelling place, the territory where she has always lived, the land where she was born, the soil that shaped her identity and gave her a sense of self. This is her island; and I am taking it away from her. I am taking it away from her because I am one of the almost 15,000 people per day every summer who disembark from the mainland.[2] Every hour at this time of year, several boats reach the harbour filled with people like me who want to flee for a while and settle temporarily on her island. Yet the marks we leave are permanent.

By escaping here, I lay hold on this woman's land, not simply because I fill up beach space, but also because the rising tourist numbers are part of the geo-dynamic processes that, together with climate change and the rising sea level, reduced sediment supply from the mainland and human development of the coastline, threaten to erode coasts and beaches in the Mediterranean. On the coastline of Porquerolles, the cliffs retreat a few centimetres per year;[3] with every new boatload of excursionists arriving at the docks, every ship of holidaymakers appearing at the port, bits of land are disappearing. Human aspirations, tourist marketing strategies, social organizations, climatic processes and geo-physical activities, all woven together; a web of dynamic encounters between CO_2 emissions, heat, travellers, holiday schedules, ships, soils and oils, waters and sand, with the result that the woman's land is vanishing.

How could I respond to her? Because of my traces, she is slowly losing her earthly belonging, leaving her devoid of space. As the result of my movements, her land is moving away from her. How could she respond to me? I am an outsider to her, a stranger, that is certain, but not a usual one. The stranger was always a character on the brink of distance and proximity, a figure *in* the group but not *of* the group.[4] By his or her unusual appearance, the stranger usually binds its members even closer to their home, but in the woman's eyes she sees a newcomer who not only intrudes upon her land, but also alters her soil and *estranges* her from her territory. She does not recognize me, but neither does she recognize her land.

My escape at the cost of the loss of her home; my freedom at the expense of her land. Like most Mediterranean islands, for hundreds of years after the Roman Empire, Porquerolles was looted by pirates who roamed the seas in search of golden treasures, dropped their anchors, raided and left. Now I have transformed from seafarer to marauder, sailing under a black flag, robbing and plundering someone else's land.

Separated by some 30 metres on the shoreline, we both feel the earth being moved, we are moved with it, and in the middle of its geological layers our two destinies collide between 'nostalgia' and 'solastalgia'. Where the former unfolds as a state of homesickness, a longing for an irrecoverable place, period or person left behind, the latter emerges in the yearning, distress or desolation caused by ongoing deprivation of the land on which you live.[5] I mourn a distanced past and a lost horizon, my purity, my freedom and the being I used to be; she the loss of her land and her soil, her homeland, as it disappears in front of her eyes.

The problems have caught up with me. They haunted me day and night in Paris, they attached to my family, and now they have found me on this island. If I ever hoped this place could offer me any solace, solitude or distance, I was wrong. There is no such perfect island and there is no isolation. Escape is but a fiction, unable to withstand the test of reality. There is simply *no outside* to these problems.

Walking in the sand along the beach on what appeared to me an island paradise, safely distanced from the issues, I suddenly have the gait of an awkward, modern pirate of the Mediterranean, involuntarily stealing a land that is not mine by my very presence. 'All hands on deck!'. . . The loss of purity, the loss of freedom, the loss of territory, the loss of place and the loss of identity, all tightly wrapped together.[6] My deprivation is ideational and hers material; the shore is moving, and the island is disappearing under our feet, as a dream and as a territory.

Freedoms

The woman's island for my freedom, my freedom for her island, but bit by bit, both seem to be dissolving. In the mirror of her despair, my freedom seems torn apart. If this island is disappearing due to the traces of my liberty, then my freedom seems to be an archaic attachment that I must abandon. Yet, despite witnessing the destruction it is wreaking on this woman's dwelling place, I still cannot let it go. Even with my liberty's purity crumbling before my eyes, I cannot abandon it. I always thought of myself as a modern human being, and, perhaps more than anything else, this meant a strong political, emotional and aesthetic attachment to freedom. I believed in it, I thought it was my horizon, and so did the society I lived in, with its social order legitimized through this ideal.[1] This was our self-understanding; imprecise, perhaps, but efficient and resonant.

Even if I stumble nervously along the coastline, the ideal and image of freedom is too strong to be a past fiction I simply leave behind on the beach. Without some formulation of freedom, I will never come to peace with my earthly anchoring, on this island or elsewhere, and I suspect that this old woman feels the same. Alone and together, our attachment to this ideal is too tightly knotted for anybody to sacrifice it on the altar of our earthly condition. 'To ecologize or to emancipate?' must not be the ultimatum, because earth does not stand a chance against the strength of freedom. I cannot practise and think of freedom the way I used to, that is certain, but neither can I renounce it altogether. I must stay *loyal* to this value, even if I have to *betray* the current meaning of it. I need to rebuild my kingdom, not tear it down.

I do believe it is possible to anchor my freedom in the soil of this island, because although my freedom today is visible in the geological layers of the shore, its meaning was never chiselled in stone. Even if some ideas of freedom gained hegemony at certain times, the understanding and experiences of this ideal have taken many forms throughout history.[2] However, given our current dominant understanding of freedom, it is not surprising that I ended up on an island, because its image mirrors its concept: in this classical liberal narrative, freedom implies the *absence* of something. Freedom is the lack of external constraints, limits or obstructions regarding my actions or movements in space. It is a freedom *from* – it has to do with keeping something at a distance.[3]

An appealing image, but no man is an island.[4] Indeed, as I learned from the old woman, not even an island is an island. The boundary between 'my body' and its 'external constraints' is not natural but artificial, and it is proving inaccurate. Microbiologists know there is no gap between bodies and external space. This is why they call us 'holobionts', a term describing how humans *consist* of the organic agents we always rendered external, but which allow us to live, breathe and move.[5] Just as the island's shore consists of the traces I leave on it, with every new step I take away from the old woman, so do I consist of the traces it leaves on me, blurring the distinction between the body organism and the external space environment. My body is not separate from the island's oxygen, the bacteria in the sand, and so on. Among other things, I *am made of* the air I breathe and the bacteria I cross paths with – in the same way as these bacteria and this air is made of myself and other entities, through time. Consequently, locating my freedom in a lack of constraints in external space leaves me in a *lack of space*, precisely because it is not situated. By disconnecting me from the entities I need to survive, and by concealing how I myself co-construct space, this idea of freedom generates an endless move beyond the limits of the earth. By making a *limit* between body and space, and by imagining freedom as an unobstructed movement through this space devoid of traces, one necessarily generates a *limitlessness*, driving one past the earthly conditions for habitability on the island. The invention of one limit – between body and space – results in limitlessness in terms of movement, with the consequence that the island's ecological limits of liveability emerge in front of our eyes.

Looking up from the sand between my feet and out over the blue sea, I realize that moving towards the open space and its endless horizon cannot make me free, because it actually makes me lose space. Freedom as an absence of relations only seems natural if I conceal the entities I live *through*, and it cannot emancipate me because it leaves me hovering in mid-air, endlessly flowing towards the destruction of these same conditions of subsistence. The sunlight glitters brightly over the waves as I walk on. These surface waves are caused by energy in the water, mostly formed by the wind, but sometimes also by animals in the water, or human activity; throwing a branch or rock into the water might create a burst of sun glitter. Where else to find my freedom if not in this horizon of external space?

Since space is altogether so troubling, it would be great to escape to an inner ideal world of morality and reason, where no problems of materiality reign. Instead of finding freedom in objective external space, I could try to locate it in my subjective internal sphere. If external nature is so out of my control, perhaps freedom and morality ought to be realized beyond its realm of necessity.[6] Maybe I could find autonomy away from the laws of nature? Facing this woman losing her soil, it seems tempting to hide inside my own inner island, but as I close my eyes and try to divorce myself from the material world, I do not see this inner citadel offering any more roads to freedom than objective external space did. What I do see is Victor, last night on the boat's deck, reminding me how one idealist philosopher once engraved his admiration for such separated 'starry skies above me, and the moral law within me'.[7] Yet, looking through the boat's porthole, the sky did not look separable from morality or freedom. This sky and these clouds full of greenhouse gases were not above my morality or liberty; the troubling forms they were taking were partly due to the free roaming of my species. Viewed from my cabin, there was no way of distinguishing between a deterministic nature and its starry sky, on the one hand, and a free, moral subject, on the other. Instead, morality and freedom entangled with nature, with the island and with the air.

The sun is about to reach its zenith, heating up the beach. It must be around noon. Feeling the midday sunrays burning my skin, I know no inner island can emancipate me, since freedom is never separate from nature. I must be able to realize my freedom *in* this world, not beyond or withdrawn from it. Continuing my hasty walk along the shore, I pass a dozen people sitting on the rocks close to the cliff wall, forming a kind of circle, eagerly discussing some political topic. Individuals trying to construct a 'we', a society. Might this be a model for my freedom, localizing freedom *in* associations with others? Communitarian philosophers thought so; for them, freedom came through good social relations and by engagement in society's ethical life. Freedom was not about keeping something at a distance or about retreating into oneself, but about ethical relations with others. Freedom was not about cutting off dependencies, but about nurturing better ones; never *outside* time and space, but always *in* society and history, with its pathway being *engagement* rather than withdrawal, its basis relational instead of purity.[8]

I have to find my freedom with others in society; it makes sense to me. The problem is that I no longer recognize the society these philosophers spoke about as it only consisted of humans. To them, freedom found its ground in the overlapping encounters between living beings, realized through their joint construction of a common ethical life. But by drawing yet another limit between 'Society' and 'Nature', they reduced this to a human enterprise, excluding numerous superimposing living entities and collectives, beings that in their entanglements with humans co-construct or destruct the ethical, living good of society. Since I arrived on this island, many other beings and lifeforms have posed me ethical questions and demanded moral actions from me. Yes, an old woman spoke, but the disappearing shores and the rising seas made her do so. Today, this island's society and ethical order is not only constituted by subjects, it is not only a sphere made of humans; it rather consists of a multitude of existences, including water, CO_2, insects, animals, forests, seagrass, air, soil, land – all demanding ethical engagement and moral reflection, in their own ways. What is more, these beings seem autonomous as well, in their specific mode of being, because *they too* make up space, dragging and borrowing traces as they go along, following their own paths, bowing to no universal laws of nature.[9]

Two boats cross each other, a stone's throw away from some kids playing on a paddleboard in the water. A few metres to their right, a plastic bottle is floating on the surface, repetitively drawn to the edge of the sand and back out again. My freedom cannot be realized through intersubjective relations. It needs to be grounded in the soil, founded with all the different sorts of earthly beings that I live with and through, negotiated with these terrestrial agents I cross on my way, that pose me moral dilemmas and demand ethical actions from me. If I want my freedom back, it must come through participation in a society of multiple entities, and by carefully engaging in reflexive negotiations with and nurturing of the diversely shaped beings that allow my continuing existence, in all its dimensions.

For my kingdom to be rebuilt and not torn down, I need to open its doors to such beings and lifeforms. To reclaim the keys to this castle, I need a new understanding of its ownership rights, since it was precisely the many trespasses over its demarcation lines that made me lose them. If I want my freedom to return, it must be through practising and cultivating links or relations with all the humans and non-humans that allow for an unfolding experience of self-determination and a sort of hetero-autonomy. This is indeed experiencing freedom ethically by 'being-myself-with-another', but it extends the scope of 'others' to include elements that I did not previously think of as part of this realm. No more limits, and hence no more boundlessness. It is by situating my freedom *inside* the earthly conditions of habitability, and in its *lack of limits* between bodies and space, between humans and non-humans, between society and nature, that my freedom avoids being *limitless*.

Landscapes

Sitting down on a rock near the cliff wall, looking towards the harbour on the horizon, I sense
the fragmentary shape of my thoughts. I know
these are only half-impressions, many intermedi-
ary calculations are necessary. Yet I cannot help
but think that if I developed, specified and followed a blueprint along these lines, perhaps I
could rebuild my freedom. However, no fallen
kingdom rises again overnight. Even if such a
reformulation of freedom is possible, developing the *feeling* of it does not come instantly. As
with all other values, it takes time, training and
nurturing to make new understandings of liberty
affectively resonate on the level of experience.
Besides, had I instantly developed such affective

registers, I would still not experience them right here and right now. Freedom is not something one cognitively realizes, but a thing one experiences through engagement in well-assembled communities, and on this island, it does not exist. If I cannot find my freedom here now, it is because this island and its collective of beings are not yet well ordered. If I remain haunted, it is not simply because my old freedom has proven impossible; it is because I am still moving through a space that is disappearing. My emancipation is possible only in a setting that does not yet exist.

My loss and the old woman's loss are not separate concerns, but two sides of the same story, as my freedom depends on an island no longer disappearing. As I stand up and turn around, I realize she is far out of sight. Lost in my hymns to freedom, I must have walked a hundred metres along the shoreline. The evidence is clear. There is no way out of this, yet I still want to escape the old woman, the rising seas and the eroding coast. Gasping for air again and almost trembling, my bodily coordination is failing me. I want to return to the harbour and to the boat, where I felt calm, but trying to find my way back,

I realize this part of the beach offers no clear paths up again. To avoid facing the old woman again, I begin climbing up the slope of the cliff, but the path ends abruptly after just a few metres. The way is blocked. On my left, a sign informs me that the authorities have cordoned off the area. Like other large parts of the island, the zone is forbidden due to the high, recurrent risk of wildfires. The sign also tells me that the cigarette I was planning to smoke to calm my nerves is illegal. Lighting it can result in a fine. Campfires and camping are forbidden too. All this because every year, vast hectares of forests, land, wildlife and vegetation remain in constant danger of bursting into flames, like the wildfires ravaging California, Greece and Australia during the summer. Each year, a bit more of the island is at risk of going up in smoke, making the air unbreathable and turning the sky red – a threat caused by low rainfall, strong winds and dry vegetation, and intensified by careless tourists, climate change and rising temperatures.

Paradoxically, the sign and the barrier sealing off the land prove it. Nature is never pure. The territory is never untouched, and wilderness never left bare; they are changed, transformed, affected and torn. Unable to pass through, I turn around and look back towards the ocean. My gaze wanders from the mainland on the other side and back to the old woman, whom I can still glimpse beneath the cliffs. Perhaps her story is more complicated than this one-sided narrative of loss. Chances are that, as one of the island's inhabitants, her economic conditions depend on me and the tourism destroying her land. Just as the problems follow me regardless of the choices I make, perhaps her life too is caught in an irresolvable riddle between survival and destruction, between gain and loss. This must be enough to render any sane soul crazy: knowing that the land belonging to you and your successors is disappearing, while being aware that this vanishing, due to the current mode of economic and territorial organization, secures your way of life.

I cannot stand to look at her, but neither can I stand to look at the ocean. One or two clouds have emerged and a group of four seagulls passes by. Trying to follow their lines of flight, I feel that my senses do not suffice. My sight seems too confined, too limited, too narrow to grasp the scheme of things, as if my gaze has constricted, as if it has shrunk, and in the process, my under-

standing of things has slowly disappeared. I have
lost sense of the land I am walking through, and,
again, the feeling of being directionless comes
back. I try to pull myself together and keep calm,
yet my vertigo returns. Standing on the edge of
the cliff, I feel trapped between sea and land,
between water and fire, with no compass to guide
me out. Maps and atlases used to lead the way,[1]
they gave us the points of reference necessary to
continue down our paths and find our route, but
as I open my bag and unfold my old map of the
island, I realize it no longer offers me any useful
coordinates. It is a 360° experience of dizziness;
no matter how I turn it, I no longer recognize
the territory in front of my eyes in its depic-
tions. If I refer back and forth between the map
and the land, the curves, connections and dead-
ends no longer end up where I expect them to.
The map gives me no realistic designation of the
territory, for it only charts the *extension* of the
land, and not its *intensity*.[2] It does not depict
the multiple, overlapping beings characterizing
this landscape's 'critical zone'[3] – the thin layer of
the earth, ranging from the trees to the ground-
water, where the interactions between rocks, soil,
water, air and living organisms have allowed Life

to flourish – and neither does it depict its loss of species, biodiversity and habitability.[4]

Atlas was the Titan who carried the heavens aloft, but they seem to have fallen from his shoulders, and crashed down in front of me. Indeed, this is what I need: a map of this crash, an Atlas who bears extinctions and disappearing places on his shoulders;[5] unfortunately Victor did not provide me with such a tool.

Turning back again, and realizing there is no other choice, I leave the sea behind me and step over the barricades shutting off the area. Dead branches crack beneath my feet as I tread forward. My steps are quick, but only a few metres into the closed zone's vegetation a voice suddenly yells at me to stop. A man appears between the trees, on the right. In a state of panic as he walks towards me, I mumble something, telling him I am lost, that I do not know where I am, and which direction to follow. Realizing that I am not trying to commit any crime, but rather that the territory has escaped my comprehension, he offers to help me find my way back to the harbour.

Waters

Laurent, a tall man in his late thirties with long hair and skin weathered by the sun, has a face of that age where youth is flowing into adulthood. He was born on the mainland, but has been living on Porquerolles for the last fifteen years. As one of the island's municipal street sweepers, he handles multiple assignments related to the cleaning and micromanagement of the roads, alleys and parks. According to him, due to his job, he knows the island's landscape better than most of the residents, including those who have been here all their lives. He is the kind of person who knows everyone, he says. Not only does his profession take him down paths and trails where others rarely walk, it also acquaints him with most of the locals, their everyday life, struggles and concerns. In other words, Laurent knows Porquerolles' territorial mosaic and social infrastructure. As we exit the closed area together, I realize I might have found someone who can guide me. Like a Virgil for Dante Alighieri, he might be able to assist me in navigating through this *terra incognita* and help me regain my sense of orientation.

Perhaps I am right, because as we start following the trails back to the harbour, Laurent immediately begins delineating the island's ecological and sociological contours. Initially, I do not confide in him my worries and confusions, but either he senses I am losing the ground underneath my feet, or else his concerns simply match mine. Whatever the case, he redirects his gaze and points towards the ocean as I did a few minutes before, and describes what I already know: namely how climate change and soaring temperatures have caused the sea levels to rise, eroding the island's coastlines and slowly swallowing the beaches. This phenomenon, he explains, is intensifying due to the decrease in *Posidonia* seagrass, which protects the beaches against erosion, and thus helps preserve the important ecosystemic roles that these sandy environments play.[1] Yet what I did not know is that these climate change-induced erosions are attacking the island's natural water source, crucial for the daily lives of its inhabitants. When the ocean rises and drowns yet another part of the shore, it threatens to contaminate and pollute the island's potable water.

A moment of silence follows as we continue walking. The worst part, Laurent suddenly sighs, is that fresh and drinkable water was already a scarce resource on the island. It barely suffices. A few times a day, a special bulk carrier tankship navigates in between the tourist boats, and arrives from the mainland with thousands of litres of additional water supply to keep the island and people from drying up. However, it is not always enough. One July day, some years ago, the island's water supply ran out, due to the mixture of a heatwave and the overload of daytrippers. There was no more potable water. One of the island's fundamental material conditions of subsistence had expired. In the same way as wildfires threaten to make forests disappear, air unbreathable and heat unbearable, and as rising temperatures jeopardize the island's agricultural crops, the heatwave mixed with mass tourism had taken away the water.

Laurent explains how the reaction that followed among the island's inhabitants and the tourists was an uncanny collective emotional register: a mixture of panic, resignation and mordant sarcasm. People were flickering between fear, surrender and self-deprecation: what kind of Mediterranean island paradise runs out of water?

I wonder whether the islanders felt some slight satisfaction seeing the tourists' fear at a water loss they themselves had triggered. Not knowing how to ask the question, I instead cut Laurent short mid-sentence, and test one of my metaphors, proposing that the island's water is being colonized. Answering yes, he mirrors the same sarcastic tone when adding how nice it could be if the occupation of water stopped there, but that unfortunately is not the case. As we reach the same gravel road I walked on a few hours earlier, a swarm of people surround us, dressed in summer clothes, on their way to the beach. Navigating in between them, Laurent details how the countless ships, tourist boats and yachts docking at the harbour contaminate the waters and pollute the seas around the island, not least through the discharge of sewage. Like many other coastal zones, he explains, the coastal waters of Porquerolles are ecologically deteriorating.[2] Despite their crystal-clear appearance, their biodiversity and ecosystems are under attack, thus the marine life that used to thrive here has died or fled elsewhere. For years, there has been a law protecting the island's waters against fishing – but according to Laurent, there are hardly any fish left. They have disappeared.

At the crossroads, another group of people crowd around us. Laurent points to a sign, proving how the tourists are on the right path to the island's most visited beach, *Plage d'argent*. This, however, is no longer the locals' name for it: due to the waters' dangerously high bacterial activity during the high season, they now refer to it as *Plage Staphylococcus*. As in a process of reverse alchemy, silver has turned into filth.

Controversies

Along the way, we pause a few times as Laurent meets and speaks to people he knows. Many of them are concerned about the ecological state of Porquerolles, not least due to the way tourism is colliding with its conditions of habitability. The disturbance of the island's biodiversity by its many visitors has long been a matter of concern, and in fact Porquerolles' cultural history is impossible to disentangle from the struggles to conserve its habitat. When selling the isle, Fournier's descendants wanted its natural landscape protected against the consequences of development, and thus rejected offers from the tourist industry, instead selling it for a lower price to the state. However, even if managed as a so-called 'Marine

Protected Area' by the Port-Cros National Park, the island has seen an enormous increase in visitors, leaving trails of destruction in their wake.

Laurent goes on to explain how this pressure on the island's landscape and biodiversity has resulted in multiple controversies among the territory's many stakeholders. Resident inhabitants, scientists, those with business interests, visitors, politicians, board directors and managers of the national park, lawyers, tourist operators and the non-human living beings of the island itself: all occupants of the island, colliding and confronting each other as they inhabit and transform the territory in different ways, with different interests and horizons, forming alliances and enmities.[1]

When I ask Laurent what these controversies are about, he points to the ground. They are all about the very land we stand on, he says, and, more specifically, the amount of visitors it can sustain. Most inhabitants of Porquerolles agree on some level of tourism: not only does the visitor flow secure certain inhabitants their salaries and the village's wealth, but people are also proud of showing off and sharing their unique island. Tourism raises awareness about protected landscapes and biodiversity, Laurent acknowledges.

However, in recent years, the ruinous ecological and social cost wrought by unrestrained development of the tourist economy has become clearer to the local community, giving rise to a new division of conflict. Today, Laurent explains, one can roughly split the island into two groups: on the one hand, those who want to maintain, develop or expand the tourist economy (often those who gain economically from it); and, on the other hand, those who want to limit the scale of tourism, and maintain or preserve the island's landscape, biodiversity and ecosystems.

As we stroll past the white and yellow stone houses I passed earlier in the morning, Laurent introduces me to one of the residents, his friend Jacob. When growing up on the island, Jacob tells me, they learned in school that a real *Porquerollais* would never litter. He and his friends were raised with what he calls an 'ecological consciousness', but now he feels his island is under attack, both from within and from without. He hates the tourists, but it is not really their fault, he says. Instead, he scorns those people in the village, mostly those associated with businesses profiting directly from an untrammelled tourist industry, who do not wish to protect the island. Some of them even claim that nothing has changed, that they see no difference – which is impossible, he says: garbage everywhere, air and water pollution, loss of wildlife and biodiversity, water shortages . . . He understands, they too need to make a living. But what about the other living beings? How can they deny the deterioration of the island? It is not only I who see these conflicts across the generations: among the people Jacob blames is his own father, who for years was one of the main actors in the development of the island's tourism.

Perhaps I was wrong, perhaps the old woman on the beach is not torn down the middle. Perhaps she has taken a side and knows where she stands in these controversies. Yet my intuition did not completely fail me: there *is* a division on this island between economization and safeguarding the island's territory by securing its ecological conditions of habitability, if not existentially then at least sociologically, thus pitching those who want to *economize* their relations to the island against those who wish to *ecologize* them. After leaving Jacob behind, Laurent tells me that since 2016, the Port-Cros National Park has tried to define a common horizon settling these disputes by setting up a collaborative study with the aim of measuring and defining Porquerolles' 'carrying capacity' in terms of visitors and tourists, one that would allow both for development and for a flourishing island.[2]

This is not just local politics; it is a politics of the land and its shape. Trying to measure such capacities is another way of asking: how is it possible to incorporate the practices of production in a wider set of processes of reproduction? When does production become counterproductive? How can production *engender* and not *endanger* the island?[3] It is indeed a way of *weighing* things differently, as the territory and its many ingredients now prove to have another shape, another set of *capacities*, that might or might not *carry* humans and non-humans alike, in more or less habitable ways. By trying to define such measures, the national park administration thus tried to mediate in controversies over how to inhabit the land. But, seemingly, they have not yet succeeded. According to Laurent, even if some restraints on tourism are imposed, they will be far from enough to secure peace; these conflicts are still proliferating.

However, what *did* provide an effective way of measuring things differently was the pandemic.

Indeed, although the tourist economy's impact on the island has been visible for years, never was the scale more clear for many of the island's inhabitants, humans as well as non-humans, than during the long period of confinement. For months, nearly no boats sailed the surrounding waters, and almost no tourists disembarked at the docks. In their absence the streets were without garbage, the air was easier to breathe, the mornings were peaceful and there were plentiful fish in the sea, insects in the soil and birds in the air. The water never ran out and the sea cleared up again, allowing a multiplicity of different beings to thrive on the island, simultaneously. Laurent recounts the day he went for a morning hike to the edge of the island with his daughter and they encountered a group of pink flamingos. What they witnessed was another *regenerative process of different lifeforms.* The island came out of hiding and its many beings crossed paths. At the same time as the island's *system of production* was paused, it became apparent how it normally functioned as a *system of destruction.*[4] The island was transforming, no longer ravaged or eroding, but now rehabilitating and regenerating. Broken links were repaired, damaged connections repaired. The disappearing island was reappearing, not because it was isolated – food, medicine, water, energy and people were still transported there every day – but because its relations to elsewhere were maintained in a *measure*, *format* or *scale* that now allowed for the reproduction of a wider set of beings than just humans. Shortly after the travel restrictions stopped, as the boats filled with tourists returned to the island, the fish disappeared once again, and the air suddenly became less breathable. Once more, the island began fading away.

What are the stakes? It is a matter of gain and loss, but of what exactly? While this island might have increased its economic growth, it has decreased in prosperity.[5] This is Laurent's isle, his co-species, his waters, his identity and his land of belonging, but the disturbing presence of other people, including myself, entangled in a tourist economy, is taking it away from him. His space is invaded; the land he lives from, the water he drinks, the air he breathes, the soil he thrives off and the numerous non-human neighbours he lives through diverse encounters with, it is all being seized. Whether Laurent speaks about the island being polluted by cars and garbage, or about its lack of parking lots, the eroding coasts or the unclean waters, it testifies to a pattern of territorial colonization. Other people are living *off* these prodigious lands, leaving their traces on the island, washing away the earthly conditions of existences permitting different forms of life to thrive here. On this island, everything that was solid seems to be melting into air.[6]

At the southern tip of the island, hidden between the waves and cliffs, lies the entrance to an underground cave where pirates used to hide their booty.[7] Today, by the same cliffs, massive swirls of garbage, plastic cans, wine bottles, food wrappers, cigarette butts and oil pools entangle together and smash against the shores, guided by ocean currents. Treasures and trash: a whole civilization later, a new kind of pirate is spreading fear, roaming and robbing the fruits of the island.

Struggles

After navigating the island's trails for about an hour, we reach the port village. It looks nothing like it did in the morning; in a matter of hours, it has transformed into a chaotic melting pot of restaurants, ice cream vendors, souvenir shops, bikes, people, voices and noises, rubbish, dust, smog, sun and heat. Overwhelmed, I look at Laurent, who sighs and shrugs his shoulders; this is nothing but an ordinary summer afternoon here.

Before moving on with his day of work, he invites me to have a drink with him, and suggests the Café Côté Port, just in front of the port's most exclusive berths. Shielding my eyes from the blinding sun, I can see, about 30 metres away, a long row of luxurious yachts and sailboats stretch-ing into the distance. For decades, Porquerolles has attracted the political, economic and cultural upper classes of France. In August, Emmanuel Macron visited his presidential residence near the island, amusing the media by cruising around the bay on a water scooter. Simultaneously, one of the world's richest men, LVMH billionaire Bernard Arnault, docked his €200,000,000, 100-metre-long luxury yacht *Symphony* and its thirty-eight-person crew at the port. At the table next to ours, novelist Delphine de Vigan is sit-ting with her family, hiding from the sun under a parasol, reading and drinking. Indeed, neither Pareto, Marx nor Bourdieu died in vain, as the political, economic and cultural class patterns are all in evidence on this island. Yet none of them offers a narrative accurate enough to describe the new *geo-social* class struggles for territory and the island's conditions of habitability unfolding here before my eyes and under my feet.[1]

Because of the island's loss of habitable territory and the struggles centred on it, I cannot simply map its classes by political power, cultural dis-tinctions or economic relations. Contaminated air and waters, the eroding coast, wildfires threatening to set the land aflame: the 'social question' entangles with the island's ecological conditions, transforming social conflicts into struggles over its earthly conditions of subsistence, and how to inhabit this land. This is what I have observed and learned here: class struggles are taking on another shape, increasingly organized around the island's territorial means of reproduction, and around ways of securing such conditions of habitability, vital to sustain different life forms.[2]

Indeed, I remember; such entities of reproduction were, in fact, part of Marx's theory on class, but they were reduced to the 'system of production' – a material sphere where classes were assembled. Yet, on this island, *production no longer seems to secure reproduction.* Instead, it seems that production partly *destroys* the continuous reproduction of the island's society and its mean of subsistence by attacking its conditions of habitability Production no longer engenders here; it endangers.

Indeed, if the problems never seem to disappear, it is because the system of production has turned into a system of devastation, reshuffling the question of the subsistence of societies, communities and their habitats. As the land disappears *because of* production, a new material basis of reproduction or engendering *beyond* production is appearing, and, with it, new class affiliations are emerging. As the island's land, soil and waters are under attack and rendered insecure, a different sort of class landscape appears: a landscape of *geo-social classes*, not assembled with regard to their relations to the means of production, but based on antagonistic relations to territorial conditions of existence, including entities such as land, climate, air, energy, soil, food, water and crops.[3]

The basis of these classes' collective interests is not their economic position but their territorial one.[4] It is an unusual alignment. Laurent, Jacob and the old woman on the beach might each have a different economic status, they certainly look as though they differ in cultural capital, but in terms of territorial affiliations, they are allied. Even if they differ slightly in their worries, they share similar relations to the land, having in common the loss of it. They overlap in their concern for their island's earthly means of subsistence, and they are engaged in the same conflicts to preserve, sustain or maintain this territory and its habitability, a struggle fought against those occupying their land, and those who wish to continue developing the tourist economy.

This struggle is not simply one of 'taking over the means of production'; rather, it looks like a struggle *for* reproduction but *against* practices of production, as these practices are making their island disappear. This is not a conflict regarding control over the forces of production, and neither is it a conflict organized around the distribution of the fruits of production; it is a struggle over the *scale* of production and a conflict regarding the *distribution* of production's destructive consequences, as it endangers the reproduction of both non-humans and humans.[5]

As I walk through the café on my way to the restroom, the TV is transmitting a speech by Jean-Luc Mélenchon. He and my Marxist friends would perhaps tell me that the exploitation of the working class has always entailed the appropriation of nature.[6] This might be true, but following Laurent's narrative, the old woman's mourning and Jacob's railing against his antagonists in the village, this is too imprecise an assessment of what is at stake on the island. Rather, new classes have assembled, based on access to and conflicts regarding habitable territories, at a moment when these are disappearing, in a way that entails a new sort of exploitation. Laurent and the old woman are not just exploited by having the surplus value of their labour taken away from them by capitalists; more dramatically, they are exploited in the sense of seeing their earthly conditions of subsistence eroding due to other people's mode of occupying their soil. Class exploitation not only unfolds *within* the production system; it has extended *beyond* it, to the sphere of reproduction, with some collectives' livelihoods upheld at the expense of other people's access to habitable territories. It no longer suffices to define exploitation by labour position and extraction of *surplus value*; exploitation is now anchored in the soil, and by the appropriation of *surplus existence* that allows some collectives' ways of life to be upheld at the cost of others' chances to maintain their basic earthly conditions of habitability.[7]

Class theory used to deliver coordinates on *what* people lived from, *where* in the chains of subsistence they were, with *whom* they shared their means of subsistence, *which* conflicts they fought and against *whom*, and *what* their temporal horizons were. This compass needs new coordinates, since Laurent, the old woman and their fellow inhabitants no longer simply live off the fruits of production, their positions are no longer inside the production system, their conflicts are no longer organized around the forces of production and their horizon is no longer that of an inevitable march towards production. On this island and beyond, a new class topography is emerging. People secure their subsistence through a wider array of earthly conditions of reproduction; they are antagonistically situated within broader processes of engendering. It is over such conditions of subsistence or habitability that social struggles unfold, and the outcome of these conflicts will define the more or less violent road that the history of humans and other beings will take.

A new climate has emerged on the Mediterranean island of Porquerolles, its territory is changing and people ally differently, fighting new pirates, facing them in a new class struggle. Here and elsewhere, the loss of land is felt under people's feet to different degrees, and in reaction, classes have assembled on the basis of their struggles to secure their territories. The earth is shaking and, with it, there is a reshuffling of the class landscape, its shape, relative hierarchies, collective interests, struggles and modes of exploitation. The pecking order has changed. In practice as well as in theory, production must give way to reproduction.

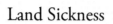

Land Sickness

Immediately after finishing his drink, Laurent insists that he has to go; he needs to visit the western part of the island. To be honest, as he disappears out of sight, I am annoyed at him for leaving me. Again, I am without a guide. I've only just started drawing up my new maps, but alone, I quickly start losing my bearings once more. Fiddling with my lighter, I impatiently survey the port. A few metres from the café, a priest is sitting on a bench, wearing a sunhat and sunglasses, eating an ice cream and tapping away on his iPhone. Perhaps he is uploading something to the Cloud, this odd new version of an immaterial-spiritual sphere, accessible by the divine grace of Silicon Valley executives. Indeed, the godly does seem a bit distant these days, perhaps because Elon Musk or Jeff Bezos has yet again departed into outer space in another attempt at transcendence . . .

A headline in the newspaper left on my table informs me of this: billionaires are floating around far above the rest of us, while down here, on this Mediterranean island, the soil is disappearing, rendering obsolete the maps we trusted to guide us along the road. Yes, our coordinates are flawed, our compasses are in need of readjustment as the magnetism of the earth is changing, but not all agree to such resets. Instead, refusing any collective responsibility for such earthly recalibrations led a few significant tech elites to draw up alternative exit maps and launch deep into space. Once upon a time, Silicon Valley was arguably a polestar for modern principles of civilizational progress, but today it seems to have become less about communal prosperity than individual survival.[1] In fact, viewed from my table, it mostly looks like an escapist fortress for the wealthy, weightless few, those owning the means of evasion. If Atlas's shoulders no longer carry the weight of the sky, and if you have the means to depart, then why not just surpass the sky itself and escape to some distant celestial body? Indeed, *Atlas Shrugged*,[2] so his admirers fuelled up their rocket ships, invested a few billion in life-prolonging research,[3] lulled those left behind into a digital metaverse and prepared to depart to infinity and beyond.

At the dawn of the Anthropocene, these stargazers know no Mediterranean island distant enough to escape these problems, and instead see no other possibility but fleeing Earth itself, together with the idea of creating a shared world. The pins on their exit maps are as territorial as they are political, and as existential as they are sociological. Realizing the soil is quivering under their feet too, they have decided to take off from all earthly limits, simultaneously cutting off all bonds of equality, solidarity and justice. Surely aware of how climate change is shrinking the planet's habitability, how there is not enough room for us all and that they too have a price to pay, these Exiters leave behind civilizational ideals of progress for all, the principles of building a habitable world for all, fleeing alone at twilight, leaving us who are deprived behind.[4]

Already in the twentieth century, psychoanalysts saw how such planetary escapes carried at their heart an existential drama: space flights to other astronomical spheres were but attempts to flee oneself, since going to Mars seems easier than penetrating one's own being, a being always confined on earth, a being who must at some point return to itself there.[5] However, now infused with the atomist venom of neoliberal ideology, this psycho-spatial spectacle has only intensified. Ultimately, this is where freedom as lack of constraints leads you: alone, floating limitlessly around in outer space, touching down on alien, uninhabited planets with no signs of life

In the event that the time-span of terraforming other planets is too vast, these evading elites have already prepared for terrestrial alternatives to the celestial heavens beyond, and marked other coordinates for their escape. Shifting the directions of escape from vertical to horizontal, from Planet B to Plan B, they have been investing in self-sufficient farmland, climate-safer survival property and luxurious disaster bunkers, in far-flung places around the world.[6] Sheltered from nuclear apocalypse, pandemic viruses or environmental calamities, these escapologists have prepared their back-up plans; safely distancing themselves from the coming floods, they remain well positioned in a climate-damaged world. In fact, it does not matter much if you colonize other planets or buy up gated, climate-secured territories, if your destination is Mars or New Zealand, they are the same kind of struggles for habitable territory, in times where Atlas has stumbled, and where such territories are disappearing. Far away from this island, perhaps, but part of the same class struggle for securing oneself a prosperous land to live off. Different strategies, positions and horizons, but in essence the same conflict over habitable space, marking the outer coordinates of a geo-social landscape.

Yes, the coast is eroding, the water is running out, the fish are fleeing, the forests are burning and the island's inhabitants mourn their loss, but on the dock newly arrived pirates are smiling while the tech billionaires are flying around Tesla cars in outer space. The land is trembling beneath all our the feet. Some are paying the costs, seeing their land vanish, while others are managing to flee – at least for a little while. As I ask the waiter for the bill, two new ships dock at the port. Another herd of tourists steps out. Girls laughing, boys chatting, fathers and mothers unfolding maps while keeping a close eye on their children. They are all like me: they are contributing to the disappearance of the island, thereby partaking in a conflict some of whose contours I might have understood, but that I am far from fully grasping.

The noise of the town, the salt in the air, it plays all around me, but I no longer pay attention. I wish I could go back to spring again, or some other kinder season. My vertigo returns, and, standing up, my knees buckle beneath me. My head is aching and my stomach is churning. I need to get back to the boat. As I am about to leave the café, Victor suddenly arrives from the harbour. He throws himself into a chair, asks me how my day was, if I found the place he spoke of, and if I like the island. Exhausted and not knowing where to begin, I tell him I feel as if the earth is moving under my feet. Undisturbed, Victor smilingly nods as if it was nothing, and goes on to explain to me that, in fact, it is a well-known phenomenon among novice sailors: the sensation of uneven ground under your feet, the feeling of the earth rocking beneath you, is often experienced among newcomers to the sea. Fighting against the wind to light up his cigarette, he tells me that I should not worry. It is completely normal, and not dangerous. I am simply suffering from 'land sickness'.

Slowly, without saying anything, I turn my head away and look towards the ocean. At the intersection of the street and the sea, the surface air looks like it is boiling. Between the asphalt and the sea, I can see waves of heat. Victor starts talking again, but his words disappear in the noise of the wind, the tourists and the seagulls. My sight blurs, but at least I am clear-minded. He is right. I am suffering from *land sickness*.

Horizons

On our way back to the boat, we pass a long line of people on the harbour waiting for what must be one of today's last boats returning to the mainland. The whole bridge is crowded, and trying to count them, I quickly lose track. Families squeezed together; people tired, but patiently waiting in line, standing shoulder to shoulder with each other. Like me, they fled for a brief moment, and now they must go home, but their footprints will last. Their trails and traces remain, affecting the inhabitants as they travel through their daily lives. Even after they leave, the island keeps carrying their weight.

Sailing towards *Timia*, Victor has to curve around a weirdly floating steel vessel coming straight at us. No one is to be seen aboard. It looks like it is floating by itself, a phantom ship without a crew. I ask Victor what this ship is and he tells me it is a sort of bulk carrier designed to convey water to the island. Turning around, I look back at the bridge. Colonists. Strangers. Settlers. Pirates. Or just ordinary people on holiday, average families with common lives, regular folks with typical jobs. They all look polite. They all look normal. They all look like I do.

As we reach the stern of *Timia*, I get the feeling that Victor is stressed. I forgot that he wanted to set sail. We cannot just stay here, he says, we need to get the boat out to sea. We need to leave straight away. There is not a minute to lose.

Stepping onto the deck, I almost slip on its wet surface, but I manage to grab the guardrail and stay on my feet. I turn around, lean over the boat's rail and gaze back at the island. From a distance, it still looks calm. Trying to locate where I was this morning, my view flies over bays and groves, from the lower parts of the beach to the upper part of the tree line, on top of the shoreline cliffs. Gazing from east to west, I catch sight of a row of luxurious coastline villas sited on the strip between the harbour and the beach, majestically facing the sea. Victor told me about them. They are the most expensive properties on the island, but they might also be the first to be swallowed up by the waves.

At the stern, Victor and Paul begin preparing *Timia* to sail. With the help of his phone, Victor first checks the wind conditions, now and to come, and scribbles them down in his notebook: 15 to 20 knots, in changing directions. With a vessel like ours and skills like his, it will be difficult but manageable, he assures me.

'Weigh anchor!' As the heavy metal spirals upwards towards the deck, leaving us floating clear of the sea floor, Paul fires up the boat's engine, directing his brother to prepare the sail. Half-prepared, half-improvising, Victor coordinates his comings and goings over the boat, from the stern to the bow. Taking hints not only from Paul, but also from the wind and the waves, and communicating back to his brother where he is in the process, Victor continually unties and reties several knots, reconsidering and reconstructing the links that will allow the wind to grab the sail, and the sail to grab the wind.

The vibrations intensify as the boat speeds up, cutting through the waves. Suddenly, Paul yells at me to take the boat's helm for a moment. I am lost, and as Paul realizes I have no clue what any of these words mean, he rephrases, and shouts that he needs me to hold the boat's 'steering wheel' straight. Following orders, I feel embarrassed for not having learned the techniques of the boat. I could not have acquired seaman skills in a day, but why was I never interested in such practicalities, techniques and materialities? I guess I always thought myself beyond or above such material ties, as if my human 'culture' alone sufficed to be in and to know the world. I never had a clue where my flat's electricity or water came from, I admit it, but, worse, I never had any interest in knowing, even if they were crucial for my daily existence. How could I not see that to understand what or where I am, I would need to extend my education?

As Paul returns to the console, he orders me to go back to the stern of the boat and stay seated. Doing as I am told, I observe Victor's coordination quickening and Paul's instructions gaining pace synchronically, corresponding to the strengthening blows of the wind, as if everything has accelerated, before it suddenly happens: in a split-second, the sail unfolds and a push simultaneously lifts the boat and drives it forward. 'Sail, done!', shouts Victor, and Paul instantly responds by shutting off the engine, leaving the rest to the wind.

Closer to the open sea, no longer sheltered by the half-moon shape of the island port's entrance, the wind-blows get wilder by the minute, but holding close to the guardrail as the swaying of the boat gets worse, I strangely feel my calm returning. I still breathe, I still move, I still pass along through life. It is as if I am somehow regaining my footing the more *Timia* is rocking, or as if the tumult of the embarkation somehow corresponds to my emotional landscape. Out here on the ocean, things seem in a way to correspond, by some means they resonate, not *despite the fact that*, but precisely *because* the boat is rocking. Here I *feel* the waves, I *see* the dangers that surround me, I experience directly the winds and the waters that threaten to throw me overboard, and whose changing moods I will have to consider carefully if I do not wish to crash, sink and drown. Here I *am* with the problems.[1]

Beware: surviving on this boat is by no means about 'controlling' or 'mastering' a surrounding

'Nature'. If we are to make it out here, we need to engage in an ongoing, discontinuous negotiation between multiple forces. The shape of the boat, its various instruments; Paul and Victor's communication, knowledge mixed with curiosity, attentiveness, precaution and imagination; their hands on rudders and sail; the shifting winds, waves and depth of the sea: all these we will have to parley with if we are to survive together. There is no such thing as 'harmony at sea', but there is more or less well-processed cooperation between all these elements, corresponding to more or less painful sailing endeavours or prosperous journeys.

Observing Paul and Victor navigating us through the harshness of the waves, I come to understand that the art of seamanship demands such negotiations, a sense of judgement and the modest acceptance of how much more there always is to learn. Sailing has nothing to do with 'progressing straight forward', which would equal certain death, as we would soon crash directly into rocky cliffs, or have our side torn apart by a violent wave. Out here on the ocean, the gravest mistake to make is to set out convinced of the direction of your voyage. To be sure, sailors do have a horizon in front of them, but it is never one of certainty, and rarely one of clarity. If the boat is not to crash, if it is to make its way through the waters and towards the horizon, it must not follow a straight line. Instead, on a boat, having a horizon always involves reinventing pathways, rediscovering the winds, repairing sails, renewing nautical strategies and perspectives as the elements change. In order to go forward, we need to continue looking back, to the sides, up, and, not least, down below our vessel.

It is a process where we have to negotiate with surprising elements and entities that move us, and that we will have to learn to move with, if our journey is not to end in tragedy. If we acquire such diplomatic competences and temporal reflexivities, *Timia* might make it through the recurring problems, through the storm we know is waiting for us, and that is already starting to blow; we might be able to go from pirates to sailors, from plundering to negotiating, from a destructive journey to a prosperous voyage. On the other side, another harbour awaits us, with another sort of earth to set foot on, a land we actually inhabit, and a soil we do not hover above; a land that is not a dream, but that might once again allow us to dream.

The sun is lower, so is the temperature, and, staring back, I notice the island almost disappearing out of sight. Below a stretch of pink, suspended in the sky, Porquerolles seems to evaporate in the distance. Turning to Victor, who is now at the console, I tell him that sailing might not be a bad metaphor for what we are facing, and that, perhaps, I will try to write about it. With a smile, he says he knows exactly what to read then, asks me to take over the helm for a moment and disappears down to the boat's cabin. As the first phrases begin taking shape in my mind, Victor returns, hands me a book and tells me that no better description exists of journeys to foreign places. There is just one thing, he adds: to understand it, I should start from the end. Politely accepting his efforts, I once again find myself flipping through sheets of paper until I arrive first at the last page, then at its final sentences:

> And [Marco] Polo said: 'The inferno of the living is not something that will be; if there is one, it is what is already here, the inferno where we live every day, that we form by being together. There are two ways to escape suffering from it. The first is easy for many: accept the inferno and become such a part of it that

you can no longer see it. The second is risky and demands constant vigilance and apprehension: seek and learn to recognize who and what, in the midst of the inferno, are not inferno, then make them endure, give them space.'[2]

Make them endure; give them space. Let them endure; let them make space . . .

Postface

Dipesh Chakrabarty

'It seems the Anthropocene is not a nice place to sleep,' writes the narrator in this book. I hover over those few words. One could not say this about other names with which we periodize human history. Could one say, for instance, that 'the feudal age is not a nice place to sleep?' One couldn't. For the word 'feudal' refers to a set of relationships conceived in the abstract. The same with 'capital'. But not so with the Anthropocene. Everything is topsy-turvy *in* the Anthropocene. Or should I say *about* the Anthropocene? The proposed name of a geological epoch, it refers to a slice of time on a geological scale; and yet it feels like a place. One could be physically 'in' it. To say that 'the Anthropocene is not a nice place to sleep' is to feel this time physically, as part of one's embodied experience. Time or – an increasingly uninhabitable – place? What, then, *is* the Anthropocene? Or does it portend a weirding of the world where time will feel like a physical place? Whatever it is, the narrator knows his present as something intimately physical. In his body. It is exhausting. One cannot sleep in the Anthropocene. The heat is unbearable.

The narrator is in Paris, a city in the grip of yet another scorching heatwave. And Paris now feels on the body like Chennai would on a hot and humid day: 'The warmth incapacitates my body and mind: everything seems slower, each minute longer, every movement heavier.' One could just as well be sweltering in the summer heat of the tropics somewhere. It is as if a European person, a young Dane to boot, experiences within himself the very distant being of the person whom the Malaysian sociologist Syed Hussein Alatas, mimicking and mocking the voice of the racist White colonizer, once called 'the lazy native'. The Anthropocene plays around with historical experience. Our young European narrator now feels like 'the lazy native'.

Seeking relief, our narrator escapes – sailing with a couple of friends – to the French island of Porquerolles. The relief he seeks is not just physical. He himself wants to be an island, shedding all the unethical connections to the non-European world he cannot avoid having, simply by dint of belonging to the Global North. He knows that the heat that makes him sleepless in Paris is itself global. This heatwave is not a 'natural disaster'. It is indissociably tied to his life, to the fossil-fuel-driven affluence of the West, to what he consumes:

> The problems fill up my basket when I do my grocery shopping at the supermarket, item by item, each one wrapped in plastic that ends up in the ocean somewhere. I stopped eating meat, but the avocados and quinoa I opted for instead cause soil degradation and water shortage where they are grown. In the mornings, the coffee I need to wake my mind destroys soils and discharges waste and pollution into foreign rivers.

The list runs on.

> Every day, I learn how another aspect of my life is interpenetrated with these problems, and how my

> being in the world entangles me in these troubles. Every day, I realize that the problem is *me*. . . . I am paying the ethical cost of my own affluence.

These thoughts have a vertiginous quality to them. Our narrator feels dizzy, nauseous. And sleepless.

Could he take responsibility for his own situation and be done with it? Could he, like Antoine Roquentin, shut himself up in some secluded place and, guided by his native Kierkegaard, dive deep into himself to achieve that Kierkegaardian ideal: to know oneself before knowing anything else? But the certainties of the old existentialism do not work for him anymore. 'It is not just that I exist *for myself*,' he writes, '. . . [r]ather, it seems that I exist *from others*, . . . some of them close, some of them far; some of them human, some of them non-human. *Intermixti, ergo sum*: I mix and interfere, therefore I am.' Or, a few lines later: 'I am earth, wind, fire and water, perhaps, but I am also soil depletion, hurricanes, wildfires and sea pollution.' He is not only connected to the rest of the world through the affluence of the West, whose scholars now make a distinction between the land you live *on* and the land you live *from*;

he also knows that the myth of the autonomous individual lies broken. The place of the indivisible and singular individual has been taken by the plural and porous holobiont, the assembly of living things one is. 'I *am made of* the air I breathe and the bacteria I cross paths with.'

The journey to Porquerolles is thus a journey of ambiguity. Does Porquerolles replace the inner island our narrator wanted to be? Or is this journey even real? Is it a dream-episode taking place within his head? The critical and admonishing voice he hears on the island – is that his own voice, the voice that chides him in Paris and keeps him awake? Whatever it is, the story is at least allegorical. Our narrator wants to get away from the hustle and bustle of Paris, from the sizzle of the heatwave, from the deeply ethical self-doubts that assail him in his sleeplessness. Maybe the air of the sea, the freshness of the landscape, and isolation from the city will help him refresh himself? But this is not to be. Not because there are tourists in their thousands crowding the beaches of the island. That is to be expected. The island is, after all, a well-known tourist destination. But there is a new problem around. Global warming has arrived here too: a 30 per cent decline in a variety of seagrass critical to the generation of oxygen speaks of the phenomenon. Rising sea waters are eroding the shores and intensifying the scarcity of fresh water that was already a problem for the island. The island now depends on importing fresh water from the mainland. And the tourists aggravate the problem. 'One July day, some years ago, the island's water supply ran out, due to the mixture of a heatwave and the overload of day-trippers. There was no more potable water. One of the island's fundamental material conditions of subsistence had expired.'

No place is spared by the Anthropocene. The topsy-turvy nature of things – even history – is visible here as well. The narrator, now a tourist, is accosted by an older woman who in no uncertain terms tells him to go. Go away, go elsewhere. He understands.

> [T]he rising tourist numbers are part of the geo-dynamic processes that, together with climate change and the rising sea level, reduced sediment supply from the mainland and human development of the coastline, threaten to erode coasts and beaches in the Mediterranean. On the coastline of Porquerolles, the cliffs retreat a few centimetres per year.

The woman – as French as any other French person, I assume – is one of the few permanent settlers on the island. But now she speaks in a voice that would remind the narrator of the Indigenous people who were robbed of their lands by the settler-colonials of Europe. It is as though the narrative of settler-colonialism is now being rehearsed – perversely – between Europeans! '[I]n the woman's eyes', the narrator is a newcomer 'who not only intrudes upon her land, but also alters her soil and *estranges* her from her territory'. This is what happened to the Indigenous peoples.

Suddenly, an onrush of historical memory. A return of 1492 and all that came in its wake. The onset of European expansion. But this is not tragedy repeating itself as farce. This is unrecognizable in older terms. Colonial practices were based on the colour line. Here the lines run every which way. Imagine a European person telling another European: don't rob me of my land. Are we all becoming Indigenous, finally? But who would be the settlers then? Those fleeing their lands because these lands have become unliveable? But aren't they also Illegals? Refugees? We don't yet have a proper name for them, though we need one.

There is no escaping the heat that burns Paris, for it is also burning other places. There is no sleepwalking our way out of this nightmare – not if one is sleepless. The narrator detects a deeper malaise. The project of the modern based on a desire to 'master' nature has failed. So even more relentless modernization could not be the way to go. There are the super-rich who dream of escaping to and surviving on another planet. Imagine: the super-rich engaging in the politics of survival, which, as a whole line of thinkers from Kant to Arendt will tell you, is no politics at all. These elitist dreams only speak of the mess modernization has made of the only habitable planet we once had. Once, we used to think that the politics of survival is what the poor engaged in. Now even the super-rich want to make it their own. A topsy-turvy world indeed!

As he makes his way back to Paris on his friends' boat, the sea becomes our narrator's road to Damascus. A vision takes hold of him. '[S]urviving on this boat', he observes,

is by no means about 'controlling' or 'mastering' a surrounding 'Nature'. If we are to make it out here, we need to engage in an ongoing, discontinuous negotiation between multiple forces. The shape of the boat, its various instruments; Paul and Victor's communication, knowledge mixed with curiosity, attentiveness, precaution and imagination; their hands on rudders and sail; the shifting winds, waves and depth of the sea: all these we will have to parley with if we are to survive together. There is no such thing as 'harmony at sea', but there is more or less well-processed cooperation between all these elements. . . . If we acquire such diplomatic competences and temporal reflexivities, *Timia* [the boat] might make it through.

If there is no project of mastery, what happens to the modern self and its pursuit of freedom(s)? Modernization has traditionally been defended on two grounds: (a) that it emancipates humans from oppression by other humans, and (b) that it liberates humans from the thraldom of nature. The trip to Porquerolles liberates our narrator from the hold of both propositions. No longer, for him, the autonomy of the Lockean self or the deep sense of responsibility of the existential one. Freedom cannot be based on any proposal of mastery of the natural world or be focused only on the master–slave dialectic between humans. Freedom has to be recalibrated.

If I want my freedom to return, it must be through practising and cultivating links or relations with all the humans and non-humans that allow for an unfolding experience of self-determination and a sort of hetero-autonomy. . . . It is by situating my freedom *inside* the earthly conditions of habitability, and in its *lack of limits* between bodies and space, between humans and non-humans, between society and nature, that my freedom avoids being *limitless*, since it is now bargained within the constraints of the earth.

* * *

In writing these lines above, I have assumed that Nikolaj Schultz, the author of this book, is not the same person as the sleepless narrator of the story we read. While the book is based on Schultz's personal experiences and encounters, it remains what he calls an 'ethnografictive' account. The story of the trip to the island of Porquerolles becomes an allegory – told in a text of hybrid genre – of a journey from sleepless anguish to some place for reflection where one human being works out their relationship to a world that is being plunged into a grave planetary crisis of anthropogenic origins. While many blame the

crisis – with good reason – on something imper-
sonal and systemic like capitalism or modernity,
Schultz begins from his own sense of personal
responsibility: the problem is *me*! This is not a
gesture of denying the larger social or sociological explanations. But it calls on us as individuals
to take responsibility for the everyday practices
that the affluent indulge in and that link us all
to the processes destroying the habitability of the
planet. How to recover the self in and after the
Anthropocene – this remarkable little book will
work like an inspiring manual for those contem-
plating that task.

Chicago
19 September 2022

Acknowledgements

This 'ethnografictive' essay draws upon my own experiences, over a few days, reconstructed as a travelogue of my journey. Based on the encounters, observations, emotions and thoughts I had, the narrative tries to unfold and rephrase certain puzzles and notions within the fields of philosophy and sociology, following the conviction that we need new analytical, stylistic and narrative approaches to do so. My deepest thanks are due to Frédérique Aït-Touati and Bruno Latour, who both read several incarnations of the manuscript, and whose comments advanced it greatly in style and thought. Exactly a month before his passing, despite his health situation, Bruno even hosted a soirée in his and Chantal's home in Paris to celebrate this book's publication in France. Words cannot describe how much I already miss my friend, his endless generosity, curiosity, warmth and wit. Among the other friends and peers who kindly commented on the essay were Ludmilla von Claer, Helle Graabæk, Marie Heinskou, Ole B. Jensen, Amalie Stahl and Lars Tønder, all of whom added their own suggestions to what it means to be suffering from land sickness. I am very grateful for their helpful interventions.

I also need to express my deep gratitude to Renaud Pacquette from Payot, not only for encouraging me to write the book, but also for his analytical and stylistic insights, which helped to give the text its shape. At Polity, I thank John Thompson for helping the English version of the book come alive, and not least Justin Dyer for helping me smooth out my awkward English. I am grateful to Dipesh Chakrabarty for his beautiful and insightful postface to this English version of the book. In addition, I need to thank Bruno Cousin and Eduardo Viveiros de Castro, whose reflections on geo-social classes helped me to fathom a bit more about this odd topic. I owe to my friends Grian Chatten, Conor Deegan III and Carlos O'Connell the insight that 'the main thing is that the rain changed direction'. And I am indebted to Mads Rasmussen and Suzanne Delatouche, who saved me from ever feeling lonely in Paris.

A grant from the French Ministry of Culture's 'Mondes Nouveaux' programme made the writing of this essay possible, and I especially need to thank Lucie Campos and Maud Desseignes for supporting the project.

A last word of gratitude is due to the people I met on Porquerolles, and to Victor and Paul: without their kindness, friendship and curiosity there would have been no book.

I dedicate this book to Vili, my godson.

Notes

Problems

1 See Cem Aydemir and Samed Ayhan Ôzsoy, 'Environmental Impact of Printing Inks and Printing Process', *Journal of Graphic Engineering and Design*, Vol. 11 (2), 2020, pp. 11–17.

2 See Emanuele Coccia, *Métamorphoses* (Paris: Bibliothèque Rivages, 2020), for an investigation of metamorphosis as a general metaphysics of being.

Beings

1 Again, this is why geologists call the era we live in the Anthropocene: the Earth System has been transformed as a consequence of human activities, and the human individual has been modified along with it, becoming a different species, a new sort of being, another kind of existence. As described in Clive

Hamilton's *Defiant Earth* (Cambridge: Polity, 2017), this novel actor wields a double-edged sword of power and impotence. On the one hand, the human species has gained an influence it never thought achievable: it now has the power to change the natural forces it was never supposed to alter. On the other hand, it seems more powerless than ever before: with nature spinning out of control, the earthly conditions humans need to survive are threatened. In other words, the new image of the human is like a pendulum, swinging back and forth between pettiness and grandeur. See, in addition, Sverre Raffnsøe, *The Human Turn* (Copenhagen: Copenhagen Business School, 2015), for many useful reflections on this new figure.

2 See Pierre Charbonnier's *Abondance et liberté: une histoire environnementale des idées politiques* (Paris: La Découverte, 2019), especially Chapter 3 on the ubiquity of the moderns and the distinction between the territory one lives *on* and the territory one lives *off*.

3 See, e.g., Carlo Ginzburg, 'Killing a Chinese Mandarin: The Moral Implications of Distance', *Critical Inquiry*, Vol. 21 (1), 1994, pp. 46–60. Chateaubriand asks the reader how they would act if they could, by a simple act of will, without leaving their room or ever being suspected or caught, kill an old mandarin living in China whose death would bring them some benefit.

4 Quote from Søren Kierkegaard's diary entries headed 'Gilleleje, August 1, 1835', as quoted in Clare

Carlisle, *Philosopher of the Heart: The Restless Life of Søren Kierkegaard* (London: Penguin Books, 2020), pp. 107–8. This gaze 'inwardly' towards existence was a fundamental trait running through Kierkegaard's work that would later influence existentialist philosophies immensely. See, e.g., the introduction to Jean-Paul Sartre, *Existentialism is a Humanism* (trans. Carol Macomber; New Haven, CT: Yale University Press, 2007).

5 While this argument is at the heart of all existentialist philosophies, it was made immortal as a phrase and slogan in Sartre's *Existentialism is a Humanism.*

6 See Bruno Latour, *The Pasteurization of France* (trans. Alan Sheridan and John Law; Cambridge, MA: Harvard University Press, 1988), for a pluralist ontological treatise, in which 'diplomacy' becomes a metaphysical principle of being, and Isabelle Stengers, *Cosmopolitics I* (trans. Robert Bononno; Minneapolis: University of Minnesota Press, 2010), and *Cosmopolitics II* (trans. Robert Bononno; Minneapolis: University of Minnesota Press, 2011), for a cosmopolitical philosophy of diplomacy.

7 See Jean-Paul Sartre, *Nausea* (trans. Robert Baldick; London: Penguin, 1965).

8 Martin Heidegger, *Being and Time* (trans. John Macquarrie and Edward Robinson; Oxford: Blackwell, 1962), Division 1, Chapter 5.

9 On the question of scalability, see Anna Lowenhaupt Tsing's work, not least *The Mushroom at the End*

of the World: On the Possibility of Life in Capitalist Ruins (Princeton: Princeton University Press, 2015), and *Friction: An Ethnography of Global Connection* (Princeton: Princeton University Press, 2005).

Generations

1 As in the classic definition of Karl Mannheim, 'The Problem of Generations', in Paul Kecskemeti (ed.), *Essays on the Sociology of Knowledge: Collected Works, Volume 5* (New York: Routledge, 1952), pp. 276–322.

2 In Bruno Latour's *After Lockdown: A Metamorphosis* (trans. Julie Rose; Cambridge: Polity, 2021), he uses Kafka's literary figure Gregor Samsa, who one morning suddenly wakes up as a beetle, to explain the metamorphoses of the New Climate Regime. See Franz Kafka, *Metamorphosis and Other Stories* (trans. Michael Hoffman; London: Penguin, 2007), and *The Trial* (trans. Idris Parry; London: Penguin, 2000).

Transmissions

1 See: https://www.worldwildlife.org/magazine/issues/spring-2014/articles/handle-with-care (t-shirt: water); https://www.worldbank.org/en/news/feature/2019/09/23/costo-moda-medio-ambiente (jeans: water); https://mycourses.aalto.fi/pluginfile.php/1240329/mod_resource/content/2/Niinima%CC%88ki%20et%20al%20%282020%29%20Environmental%20price%20of%20Fashion%20%5BNatRev%5D.pdf (jeans: water & CO_2); https://

news.mit.edu/2013/footwear-carbon-footprint-0522 (trainers: CO_2).

2 See Patrice Maniglier, *Le philosophe, la terre et le virus* (Paris: Les Liens qui Libèrent, 2021), especially Chapter 1, for a Science and Technology Studies analysis of COVID-19. See also Frédérique Aït-Touati, '"Nous ne sommes pas le nombre que nous croyions être": Relire *Les Microbes* de Bruno Latour', *AOC Media*, 23 March 2020, https://aoc.media/critique/20 20/03/22/nous-ne-sommes-pas-le-nombre-que-nous -croyions-etre-relire-les-microbes-guerre-et-paix-de -bruno-latour/.

3 Or, in French, 'J'avance masqué.'

4 On the metaphor of contagion in the human and social sciences, see Peta Mitchell, *Contagious Metaphor* (London: Bloomsbury Academic, 2012).

Oceans

1 Jules Verne, *Twenty Thousand Leagues under the Seas* (trans. William Butcher; Oxford: Oxford University Press, 2019). See also Elena Past, 'Island Hopping, Liquid Materiality, and the Mediterranean Cinema of Emanuele Crialese', *Ecozon@*, Vol. 4 (2), 2013, pp. 49–66.

2 See Carl Schmitt, *Land and Sea: A World-Historical Meditation* (trans. Samuel Garrett Zeitlin; Candor, NY: Telos Press, 2015).

3 https://www.wwf.fr/sites/default/files/doc-2017-09

/170927_rapport_reviving_mediterranean_sea_econ omy.pdf.

4 To paraphrase Camille Schmoll, *Les damnées de la mer: femmes et frontières en Méditerranée* (Paris: La Découverte, 2020).

5 On the conflicting images of the 'global' and the 'planetary' and the historical relation between them, see Dipesh Chakrabarty, *The Climate of History in a Planetary Age* (Chicago: University of Chicago Press, 2021).

6 For introductions to the notion of 'planetary boundaries', see, e.g., Johan Rockström et al., 'Planetary Boundaries: Exploring the Safe Operating Space for Humanity', *Ecology and Society*, Vol. 14 (2), 2009, art. 32, and Will Steffen et al., 'Planetary Boundaries: Guiding Human Development on a Changing Planet', *Science*, Vol. 347 (6223).

Islands

1 See William Luret's biographical fiction on Fournier, *L'homme de Porquerolles* (Paris: JC Lattès, 1996).

2 Vincent Vlès, 'Construction partagée d'un système numérisé de gestion des capacités de charge touristique du Parc national de Port-Cros', rapport final au Parc national de Port-Cros, 2018, https://hal.archives -ouvertes.fr/hal-01968691/document.

3 Georges Rovéra et al., 'Preliminary Quantification of the Erosion of Sandy-Gravelly Cliffs on Porquerolles Island (Provence, France) through

Dendrogeomorphology, Using Exposed Roots of Aleppo Pine (*Pinus halepensis Mill.*)', *Geografia Fisica e Dinamica Quaternaria*, Vol. 36 (1), 2013, pp. 181–7.

4 See Georg Simmel, 'The Stranger', in Georg Simmel, *On Individuality and Social Forms* (ed. Donald N. Levine; Chicago: University of Chicago Press, 1971), pp. 143–50 and Margaret Mary Wood, *The Stranger: A Study in Social Relationships* (New York: Columbia University Press, 1934).

5 See Glenn Albrecht, '"Solastalgia": A New Concept in Human Health and Identity', *Pan (Philosophy, Activism and Nature)*, Vol. 3, 2005, pp. 41–55, and Glenn Albrecht et al., 'Solastalgia: The Distress Caused by Environmental Change', *Australasian Psychiatry*, Vol. 15, 2007, pp. 95–8.

6 Rebecca Elliott, 'The Sociology of Climate Change as a Sociology of Loss', *European Journal of Sociology*, Vol. 59 (3), 2018, pp. 301–37.

Freedoms

1 On social order, legitimation and freedom, see, e.g., Axel Honneth, *Freedom's Right: The Social Foundations of Democratic Life* (trans. Joseph Ganahl; New York: Columbia University Press, 2014). See also Mikael Carleheden and Nikolaj Schultz, 'The Ideal of Freedom in the Anthropocene: A New Crisis of Legitimation and the Brutalization of Geo-Social Conflicts', *Thesis Eleven*, Vol. 170 (1), 2022, pp. 99–116, where many arguments from this chapter were previously outlined.

2 See Isaiah Berlin, 'Two Concepts of Liberty', in Isaiah Berlin, *Four Essays on Liberty* (Oxford: Oxford University Press, 1969), pp. 166–217.

3 The canonical example of this 'negative' understanding of freedom, to use Isaiah Berlin's terminology (ibid.), is found in Thomas Hobbes, *Leviathan* (ed. Edwin Curley; Indianapolis: Hackett, 1994). See also Philip Pettit, 'Liberty and Leviathan', *Politics, Philosophy and Economics*, Vol. 4 (1), 2005, pp. 131–51.

4 As John Donne, a seventeenth-century contemporary of Hobbes, put it. See John Donne, *No Man is an Island: A Selection from the Prose of John Donne* (ed. Rivers Scott; London: Folio Society, 1998).

5 The concept of 'holobionts' was developed in Lynn Margulis and René Fester (eds), *Symbiosis as a Source of Evolutionary Innovation: Speciation and Morphogenesis* (Cambridge, MA: MIT Press, 1991). See also Lynn Margulis and Dorion Sagan, *Microcosmos: Four Billion Years of Evolution from our Microbial Ancestors* (Berkeley: University of California Press, 1997), and Scott F. Gilbert, Jan Sapp and Alfred I. Tauber, 'A Symbiotic View of Life: We Have Never Been Individuals', *The Quarterly Review of Biology*, Vol. 87 (4), 2012, pp. 325–41.

6 A conception of freedom defended by Immanuel Kant. Isaiah Berlin refers to this as 'positive freedom as self-direction', while Frederick Neuhouser calls it 'moral freedom'. See Immanuel Kant, *Groundwork of the Metaphysics of Morals* (trans. and ed. Mary

Gregor; Cambridge: Cambridge University Press, 1998), and *Critique of Practical Reason* (trans. and ed. Mary Gregor; Cambridge: Cambridge University Press, 1997). See also, again, Berlin, 'Two Concepts of Liberty', and Frederick Neuhouser, *Foundations of Hegel's Social Theory* (Cambridge, MA: Harvard University Press, 2000).

7 The quote stems from Kant, *Critique of Practical Reason*, and was engraved on his tombstone. See Paul Guyer, 'Introduction: The Starry Heavens and the Moral Law', in Paul Guyer (ed.), *The Cambridge Companion to Kant* (Cambridge: Cambridge University Press, 1992), pp. 1–25.

8 A concept of freedom outlined in Georg Wilhelm Friedrich Hegel, *Philosophy of Right* (trans. S.W. Dyde; New York: Cosimo, 2008). Isaiah Berlin ('Two Concepts of Liberty') calls this a 'positive freedom as self-realization', while Frederick Neuhouser (*Foundations of Hegel's Social Theory*) terms it 'social freedom'. More recently, this idea of freedom has been reinvigorated as the cornerstone of Axel Honneth's social philosophy. See, e.g., Honneth, *Freedom's Right*.

9 Bruno Latour and Timothy Lenton, 'Extending the Domain of Freedom, or Why Gaia is So Hard to Understand', *Critical Inquiry*, Vol. 45 (3), 2019, pp. 659–80.

Landscapes

1 See Denis Cosgrove (ed.), *Mappings* (Chicago: University of Chicago Press, 1999), and *Apollo's Eye: A Cartographic Genealogy of the Earth in the Western Imagination* (Baltimore, MD: Johns Hopkins University Press, 2001).

2 See Frédérique Aït-Touati, Alexandra Arènes and Axelle Grégoire, *Terra Forma: Manuel de Cartographies Potentielles* (Paris: Éditions B42, 2019). For a more general cartography of the Anthropocene, see François Gemenne et al., *Atlas de l'Anthropocène* (2nd edition; Paris: Presses de Sciences Po, 2021), and Anna L. Tsing et al. (eds), *Feral Atlas: The More-Than-Human Anthropocene* (Stanford, CA: Stanford University Press, 2021), https://feralatlas.org/.

3 See, e.g., Jérôme Gaillardet, 'The Critical Zone, a Buffer Zone, the Human Habitat', in Bruno Latour and Peter Weibel (eds), *Critical Zones: The Science and Politics of Landing on Earth* (Cambridge, MA: MIT Press, 2020), p. 122, and Alexandra Arènes, 'Traveling through the Critical Zone', in ibid., pp. 130–5.

4 For the hypothesis that a global 'sixth mass extinction' of species is currently underway, see Gerardo Ceballos et al., 'Accelerated Modern Human-Induced Species Losses: Entering the Sixth Mass Extinction', *Science Advances*, Vol. 1 (5), 2015, pp. 1–5, and Gerardo Ceballos, Paul R. Ehrlich and Rodolfo Dirzo, 'Biological Annihilation via the Ongoing Sixth Mass Extinction Signaled

by Vertebrate Population Losses and Declines', *Proceedings of the National Academy of Sciences*, Vol. 114 (30), 2017, pp. 6089–96.

5 As, for example, pursued in Christina Conklin and Marina Psaros, *The Atlas of Disappearing Places: Our Coasts and Oceans in the Climate Crisis* (New York: The New Press, 2021).

Waters

1 Charles-François Boudouresque et al., 'The High Heritage Value of the Mediterranean Sandy Beaches, with a Particular Focus on the *Posidonia oceanica* "banquettes": A Review', *Scientific Reports of Port-Cros National Park*, 31, 2017, pp. 23–70.

2 See, e.g., Marc L. Miller and Jan Auyong, 'Coastal Zone Tourism: A Potent Force Affecting Environment and Society', *Marine Policy*, Vol. 15 (2), 1991, pp. 75–99.

Controversies

1 See Anne Cadoret, 'Conflicts and Acceptability of Visitation Management Measures for a Marine Protected Area: The Case of Porquerolles, Port-Cros National Park', *Ocean and Coastal Management*, Vol. 204, 2021, art. 105547.

2 See, e.g., Valérie Deldrève and Charlotte Michel, 'The Carrying Capacity Approach on Porquerolles (Provence, Port-Cros National Park, France): From Prospective to Action Plan', *Scientific Reports of the*

Port-Cros National Park (Parc National de Port-Cros, 2019), pp. 63–100.

3 On the concept of 'engendering', see especially Latour, *After Lockdown.*

4 The analysis of this dialectic was a recurring theme in the oeuvre of Ulrich Beck. See, e.g., *Risk Society: Towards a New Modernity* (trans. Mark Ritter; London: SAGE Publications, 1992), *World Risk Society* (Cambridge: Polity, 1998) and *Metamorphosis of the World: How Climate Change is Transforming Our Concept of the World* (Cambridge: Polity, 2016).

5 On the necessity of moving from 'growth' to 'prosperity', see Bruno Latour and Nikolaj Schultz, *Mémo sur la nouvelle classe écologique: comment faire émerger une classe écologique consciente et fière d'elle-même* (Paris: La Découverte, 2022).

6 Karl Marx and Friedrich Engels, *The Communist Manifesto* (London: Penguin Books, 2011), p. 68.

7 Louise Roche and Jules Masson Mourey, 'Le trou du pirate: prospection inventaire et relevés topographiques en région Provence-Alpes-Côte d'Azur sur l'île de Porquerolles – commune d'Hyères (Var, 83)', Rapport d'opération SRA PACA (Rapport de recherche), RAP07078 (EHESS–Paris; Aix-Marseille Université, 2014), ffhal-01441755f, https://hal.archives-ouvertes.fr/hal-01441755/document.

Struggles

1 See Jakob Stein Pedersen, Bruno Latour and Nikolaj Schultz, 'A Conversation with Bruno Latour and Nikolaj Schultz: Reassembling the Geo-Social', *Theory, Culture & Society*, Vol. 36 (7–8), 2019, pp. 215–30, and Nikolaj Schultz, 'New Climate, New Class Struggles', in Latour and Weibel (eds), *Critical Zones*, pp. 308–11.

2 On environmental and territorial inequalities, see, e.g., Lucas Chancel, *Insoutenables inégalités: pour une justice sociale et environnementale* (Paris: Les Petits Matins, 2017). See also Ulrich Beck, 'Remapping Social Inequalities in an Age of Climate Change: For a Cosmopolitan Renewal of Sociology', *Global Networks*, Vol. 10 (2), pp. 165–81.

3 See also Nikolaj Schultz, 'Did the Pandemic Teach Us Something New about Class? COVID-19 as an Experiment in "Geo-Social Class" Interests', in Michael J. Ryan (ed.), *COVID-19, Volume 3: Cultural and Institutional Changes and Challenges* (London: Routledge, 2022), Chapter 5 (forthcoming).

4 See Karl Polanyi, *The Great Transformation: The Political and Economic Origins of Our Time* (Boston: Beacon Press, 2001), for a precursory understanding of geo-social classes.

5 See again Latour and Schultz, *Mémo sur la nouvelle classe écologique.*

6 See, e.g., Jason Moore, *Capitalism in the Web of Life: Ecology and the Accumulation of Capital* (London:

Verso, 2017). For a critique of the eco-Marxist perspective on class, see Nikolaj Schultz, 'Ecology and Class: Three Ways of Theorizing Class in a Time of Global Climate Change' (forthcoming).

7 Nikolaj Schultz, 'Geo-Social Classes', in Marianne Krogh (ed.), *Connectedness: An Incomplete Encyclopedia of the Anthropocene* (Copenhagen: Strandberg Publishing, 2020), pp. 204–7.

Land Sickness

This chapter draws on Nikolaj Schultz, 'La vie comme exode', *Terrestres: Revue des livres, des idées et des écologies*, 16 May 2022, https://www.terrestres.org/2022/05/16/la -vie-comme-exode/, and 'At tage til Mars er ikke progressivt, det er reaktionært', *Atlas Magasin*, 2 February 2020, https://atlasmag.dk/samfund/tage-til-mars-er-ikke-progressivt-det-er-reaktion%C3%A6rt.

1 Douglas Rushkoff, 'Survival of the Richest', *Medium*, 5 July 2018, https://www.onezero.medium.com/ survival-of-the-richest-9ef6cddd0cc1. For essential clues to understand the ideology of the tech-escapists, see James Dale Davidson and William Rees-Mogg, *The Sovereign Individual: How to Survive and Thrive during the Collapse of the Welfare State* (New York: Simon & Schuster, 1997).

2 Ayn Rand, *Atlas Shrugged* (New York: Signet, 1957).

3 Jessica Powell, 'The Rich Will Outlive Us All', *Medium*, 3 January 2018, https://medium.com/s/20

69/what-happens-when-the-rich-live-decades-longer
-than-the-rest-of-us-2dfec4a35b21.

4 See the 'political fictive' hypothesis in the first chapters of Bruno Latour, *Down to Earth: Politics in the New Climatic Regime* (trans. Catherine Porter; Cambridge: Polity, 2018).

5 Carl Gustav Jung, *C. G. Jung Speaking: Interviews and Encounters* (eds William McGuire and R. F. C. Hull; Princeton: Princeton University Press, 1977), p. 468.

6 See, e.g., Olivia Carville, 'The Super Rich of Silicon Valley Have a Doomsday Escape Plan', *Bloomberg*, 5 September 2018, https://www.bloomberg.com /features/2018-rich-new-zealand-doomsday-preppers/; Jim Dobson, 'The Shocking Doomsday Maps of the World and the Billionaire Escape Plans', *Forbes*, 10 June 2017, https://www.forbes.com/sites/jimdobson/2017/06/10/the-shocking-doomsday-maps-of-the -world-and-the-billionaire-escape-plans/?sh=189d700 e404; and Mark O'Connell, *Notes from an Apocalypse* (New York: Doubleday, 2020).

Horizons

1 See Donna Haraway, *Staying with the Trouble: Making Kin in the Chthulucene* (Durham, NC: Duke University Press, 2016).

2 Italo Calvino, *Invisible Cities* (trans. William Weaver; San Diego, CA: Harcourt Brace & Company, 1974), p. 165.

Milton Keynes UK
Ingram Content Group UK Ltd.
UKHW050800090823
426298UK00003B/1

9 781509 556120